Vesselin Petkov

Inertia and Gravitation

From Aristotle's Natural Motion to Geodesic Worldlines in Curved Spacetime

New Printing

MINKOWSKI
Institute Press

Vesselin Petkov
Institute for Foundational Studies Hermann Minkowski
Montreal, Quebec, Canada
vpetkov@minkowskiinstitute.org
http://minkowskiinstitute.org/

Cover: based on NASA images

ISBN: 978-0-9879871-6-7 (ebook)
ISBN: 978-0-9879871-5-0 (softcover)

Minkowski Institute Press
Montreal, Quebec, Canada
http://minkowskiinstitute.org/mip/

For information on all Minkowski Institute Press publications visit our
website at http://minkowskiinstitute.org/mip/books/

Dedicated to the memory of Hermann Minkowski

Had he lived longer he would have almost certainly noticed that inertia could be regarded as arising from a four-dimensional stress in the deformed worldtube of an accelerating particle and therefore inertia would turn out to be another manifestation (along with length contraction) of the four-dimensionality of the absolute world of his spacetime formulation of special relativity.

Preface

There are two major reasons for writing this book.

1. To fill a gap in the literature. So far there has been no book which deals with inertia and gravitation by explicitly addressing open questions and issues which have been hampering the proper understanding of these phenomena. This book places a strong emphasis on the physical understanding of the main aspects and features of inertia and gravitation. It discusses questions such as:

- Are inertial forces fictitious or real?

- Is there a difference between active and inertial forces?

- Does Minkowski's four-dimensional formulation of special relativity provide an insight into the origin of inertia?

- Does mass increase relativistically?

- Had it been possible to realize before the advent of general relativity that there were serious problems with Newton's view of gravity as a force?

- As gravity is not a force in general relativity what is the nature of the force of weight, which has been traditionally regarded as gravitational force?

- Why are inertial forces and forces of weight equivalent?

- Why is the inertial mass equivalent to the (passive) gravitational mass?

- Are gravitational phenomena caused by gravitational interaction according to general relativity?

- Is there gravitational energy?

- Do gravitational waves carry gravitational energy?

- Can gravity be quantized?

2. To demonstrate through analyses of key features of inertia and gravitation how an innovative research strategy works. This strategy,

which is based on the most successful methods behind the greatest discoveries in physics, is offered as a way out of the present breakthroughless situation in fundamental physics, which I think is caused not by the lack of talented physicists, but rather by problematic general views on how one should do physics. The proposed strategy will be employed at a new research institute – the *Institute for Foundational Studies "Hermann Minkowski"* (http://minkowskiinstitute.org/) in Montreal, Canada.

This book is specifically written for the future physicists, who are now learning the art of doing physics, with the hope that it may help them become innovative and productive researchers. However, the technical level of the book makes it possible for anyone who is interested in the nature of inertia and gravitation to understand almost all of the issues discussed in the book.

Montreal, December 17, 2012   *Vesselin Petkov*

October 19, 2019: new printing – an Appendix "On Relativistic Mass" is included, some notes and references are added and noticed typos are corrected.

Contents

1. Introduction

Inertia is a complex phenomenon which involves two aspects. The first aspect was extensively studied by Galileo and captured in Newton's first law of motion. It deals with the fundamental property of every *free* material particle to preserve its inertial state, i.e. its state of motion with constant velocity (including its state of rest since it corresponds to a constant velocity = zero); this aspect is often expressed by simply saying "a particle moves by inertia." The second aspect of inertia reflects the *resistance* a particle offers when prevented from moving by inertia, i.e. when the particle is subjected to a force, which changes its velocity and accelerates it. The measure of the resistance a particle offers to its acceleration is called the particle's inertial mass.

Gravitation is perhaps an even more complex phenomenon. It refers to another fundamental property of particles – to attract one another. A particle under the influence of gravitation *accelerates* which led Newton to postulate that gravitation manifests itself through gravitational forces (causing the gravitational acceleration). Two masses are defined in the case of gravitation. The active gravitational mass is the source of gravitational attraction, whereas the passive gravitational mass applies to particles subject to gravitation and is defined (in the Newtonian gravitational theory) in the same way as the inertial mass – as the resistance a particle offers to its acceleration caused by a gravitational force.

Inertia and gravitation have been outstanding puzzles for centuries. The origin of inertia (and therefore of mass) continues to be an open question in the beginning of the 21st century. Einstein's general relativity provided a revolutionary insight into the nature of gravitation by regarding it as a manifestation of the curvature of spacetime, but brought about one major open question – how matter curves spacetime.

The purpose of this book is to discuss in detail the nature of inertia and gravitation with a special emphasis on drawing enlightening insights into these phenomena from special and general relativity.

As the main focus of this book is on the understanding of inertia and gravitation, it is necessary from the very beginning to state how I think genuine and deep understanding of physical phenomena can be achieved. Clarifying this issue is of crucial importance not only for the understanding of inertia and gravitation but for the advancement of fundamental physics in general. The reason is that persistent misconceptions on how one should do physics (to *understand* physical phenomena, not just describe them) might have been responsible for the fact that for several decades there has been no breakthrough in fundamental physics as revolutionary as relativity and quantum physics despite the unprecedented advancements of applied physics and technology. The fact that for decades the efforts of so many brilliant physicists to make a major discovery have not been successful seems to indicate that those efforts might not have been in the right direction.

A factor which might have contributed to this situation is that in the second half of the twentieth century teaching physics, I think, took a wrong turn in a number of countries, especially in North America. Most probably, due to the continuing difficulties to understand quantum phenomena, explanations in teaching physics are often substituted with calculations and students are encouraged and sometimes even urged (Shut up and calculate![1]) to do the same[2]. However, unlike in mathematics, in physics calculations play a *secondary* role (correct calculations are a necessary, but not a sufficient condition for a theory to be correct); the ultimate goal of a physical theory is to provide us with understanding of the physical phenomena represented by the theory. It should be stressed that genuine understanding is needed not only for satisfying our intellectual curiosity, but it is the major necessary condition for the advancement of science – only proper *understanding* of the world can prepare the ground for the next breakthrough.

Whenever possible students should be given the appropriate physical

---

[1] Attributed to N. David Mermin (Physics Today, April 1989, page 9), see also http://fisica.ciencias.uchile.cl/~emenendez/uploads/Cursos/callate-y-calcula.pdf.

[2] In fact, the difficulties to comprehend the strange quantum world are only the most recent justification for further shifting the focus in teaching physics away from explanations. Although no discovery is possible without good understanding (based on some kind of explanation) of the physical phenomena behind it, providing explanations of phenomena have always been notoriously difficult in physics. In addition to quantum physics there is another famous example from the history of physics which demonstrates that at a given stage physics can successfully advance by providing just the mathematical description of the observed phenomena and by avoiding any detailed explanations *at that stage* – the discovery of the law of gravitation by Newton [2]: "Without entering into the question that was uppermost in the minds of his contemporaries – the question as to *why* bodies attract each other – Newton was satisfied with formulating the specific connection of the attractive force between two bodies with their masses and with the distance between them...without aiming at an explanation of phenomena."

picture (by honestly stating the difficulties, the limitations, and the open questions) in order that they have an idea of *what* they calculate. I think what Wheeler called first moral principle captures the essence of successful teaching of physics and gives an example of how that can be achieved when counter-intuitive subjects are taught [1]:

> Never make a calculation until you know the answer. Make an estimate before every calculation, try a simple physical argument (symmetry! invariance! conservation!) before every derivation, guess the answer to every paradox and puzzle. Courage: No one else needs to know what the guess is. Therefore make it quickly, by instinct. A right guess reinforces this instinct. A wrong guess brings the refreshment of surprise.

To change the situation in fundamental physics not only should teaching physics be designed with an emphasis on understanding the physical foundations of the studied phenomena, but should also help the students learn the basic components of an innovative research strategy, which makes understanding an explicit priority, and which will later make them productive researchers. A scientific discovery made by the experimental confirmation of the predictions of a theory (as was the case with general relativity) is the best proof that we understand the world. For this reason. a research strategy which can lead to deep understanding of physical phenomena should be based on a thorough analysis of the groundbreaking discoveries in physics aimed at decoding how a discovery is made (i.e. how a more detailed understanding of the world is achieved). The analyses in this book employ such a strategy[3] not only to arrive at a deeper understanding of inertia and gravitation but also to demonstrate its effectiveness[4]. This strategy, which identifies, synthesizes, and develops the successful methods employed by the

---

[3]This research strategy, which has been developed for over fifteen years, incorporates and expands the essence of the successful analyses, methods, and techniques employed by great scientists such as Galileo Galilei, Isaac Newton, Albert Einstein, Hermann Minkowski, Louis de Broglie, Paul Dirac and the other founders of quantum mechanics. The main components of this research strategy form a one-semester course "Exploring the internal logic of fundamental ideas: shorter paths to scientific discoveries." As indicated in the Preface this strategy will be put into action at the new Minkowski Institute and one of its goals is to show that an innovative research strategy is by far more important for making discoveries in physics than investing huge funds.

[4]I have no intention to try to prove (especially in such a small book) why an innovative research strategy extracted from the successful approaches that led to the most important results in physics, like the one outlined here and used in this book, should contain the key elements behind the revolutions in physics, and why such a strategy should be adopted (or at least considered) by all who do science or

4

greatest physicists, consists of two main components – an adequate view of the nature of physical theories and thorough conceptual analyses so successfully involved in all great discoveries in physics.

1.1 An adequate view of the nature of physical theories is a necessary condition for making discoveries in physics

Many physicists may probably tend to regard the statement in the section's title as belonging to philosophy, which should have no place in physics. The reason is that physicists generally think they do not need any philosophical position for their research. But despite that tendency physicists do hold implicit or explicit views on the nature of physical theories[5], which vary widely – from (i) believing that our theories capture objective physical laws and that the essential elements of the theories have counterparts in the physical world, to (ii) insisting that the most successful abstractions of our theories (whose introduction usually resolve open questions) do not (necessarily) reflect real properties of the world (i.e. that our abstractions do not have analogs in the world),[6] and to (iii) even such extreme views that physics tells us nothing about the *existence* of what it describes.

A careful examination of the greatest discoveries in physics unequivocally demonstrates that the correct view of the nature of physical theories is (i). In fact, it is not even necessary to study the history of physics to see why this is so. It is sufficient to look at just a single project of the research at the Large Hadron Collider, for instance – the hunt for the Higgs boson. The effort was to verify whether the theoretical entity "Higgs boson" had a counterpart in the physical world. This is an obvious and irrefutable example that the essential elements of the physical theories should correspond to real objects or properties of the

---

would like to understand science and the physical world. I think, as always in similar situations, it is the results obtained by the employment of such a research strategy that will do the proving and the convincing.

[5]In this sense Daniel Dennett may have a good point [3]: "Scientists sometimes deceive themselves into thinking that philosophical ideas are only, at best, decorations or parasitic commentaries on the hard, objective triumphs of science, and that they themselves are immune to the confusions that philosophers devote their lives to dissolving. But there is no such thing as philosophy-free science; there is only science whose philosophical baggage is taken on board without examination."

[6]Recently Mermin explicitly urged physicists to adopt that view: "It is a bad habit of physicists to take their most successful abstractions to be real properties of our world" [4, p. 8]

world. Because of this our theories provide us with reliable knowledge and therefore with *true understanding* of those areas of the world which we study.

The fascinating and challenging art of doing physics is to identify which theoretical entities of a physical theory are essential and should have counterparts in the world and which are the pieces of reliable knowledge about the studied part of the world (which will be preserved in any future theory) that are contained in the theory. I would like to stress as forcefully as possible that such examinations of physical theories are *not* philosophy as some physicists appear to think. By thoroughly analyzing the structure and the mathematical formalism of a physical theory theoretical physicists and mathematicians not only can identify the elements of reliable knowledge in it, but can also detect its limitations and thus outline the direction in which a new theory should be sought. An example of how such analyses can be carried out was given by Hermann Minkowski[7] in his groundbreaking paper "Space and Time" [6] where he pointed out that the limitations of classical mechanics might have been discovered by "a mathematician with a free imagination". Minkowski examined the two-fold invariance of the equations of Newtonian mechanics (an arbitrary change of position of a system and a change of its state of motion by imparting to it any uniform translation) and noticed that it can be formally represented by two groups of transformations for the differential equations of mechanics G_c and G_∞ (one containing a positive parameter c, which can be interpreted as a limiting velocity in Nature, and the other without such a parameter, which means that an object can in principle travel with an infinite velocity as is the case in the Newtonian mechanics) [6, pp. 111-112]:

> By taking this into account it becomes clear that the group G_c in the limit $c = \infty$, that is the group G_∞, is exactly the complete group which is associated with the Newtonian mechanics. In this situation, and since G_c is mathe-

---

[7]Minkowski gave an example of the predictive power of mathematics, which is unfortunately so easy to be abused. In order to avoid such abuses (which can be recognized by the lack of fundamental physical results), one should acquire deep understanding of the "unreasonable effectiveness of mathematics in the natural sciences" (as Wigner [5] put it). It will make it possible to arrive at discoveries in physics by exploring the internal logic of mathematical ideas, because "mathematics has the tremendous creative power of evolving hidden consequences from assumptions about the observable universe and thus prompting predictions of previously unobserved phenomena. Not that the universe is under any obligation to conform to those predictions! But if these consequences are verified, then mathematics has led to new discoveries; and if they are not borne out by observations, then mathematics has necessitated a revision of the underlying general assumptions" [2].

6

matically more understandable than G_∞, there could have probably been a mathematician with a free imagination who could have come up with the idea that at the end natural phenomena do not actually possess an invariance with the group G_∞, but rather with a group G_c with a certain finite c, which is *extremely great* only in the ordinary units of measurement. Such an insight would have been an extraordinary triumph for pure mathematics. Now mathematics expressed only staircase wit here, but it has the satisfaction that, due to its happy antecedents with their senses sharpened by their free and penetrating imagination, it can grasp the profound consequences of such remodelling of our view of nature.

Despite the overwhelming examples demonstrating that our physical theories, whose predictions have been repeatedly confirmed by experiment, do explain how the studied part of the world works (that is why they make correct predictions), inexplicably there exist misconceptions about the nature of physical theories like the ones mentioned above. Potentially most damaging for the advancement of fundamental physics are inadequate views on how one should treat different theories that describe the same physical phenomena. Often one can hear or read that such theories are equivalent since they are just different descriptions of the phenomena. In some cases such a position is fine, but in other cases it can prevent physicists from seeing the real problems that should be addressed.

Again, the art of doing physics makes it also possible to determine whether different theories are indeed simply different descriptions of the same physical phenomena (as is the case with the three representations of classical mechanics – Newtonian, Lagrangian, and Hamiltonian), or *only one* of the theories competing to describe and explain given physical phenomena is the correct one (as is the case with general relativity, which identifies gravity with the non-Euclidean geometry of spacetime, and other competing theories, which regard gravity as a force).[8]

Without further examining the second and third view listed in the beginning of this section, I think it is more helpful to give the following two examples from the history of the twentieth century physics that

---

[8]The difference between these two cases can be illustrated by an example from everyday life. The first case – different theories are just different descriptions of a given phenomenon – is like the description of an event in different languages; every language correctly describes the event and therefore the different languages' accounts are equivalent. The second case – only one of the theories describing a given phenomenon is correct – is like different accounts of the same event (in any language) and, obviously, only one is correct.

demonstrate that these are confusing and unproductive views, which negatively affect not only the advancement of physics, but also the physicists who share them and who often pay a high personal price.

It was precisely the view that the successful abstractions in the physical theories should not (necessarily) be regarded as representing something real, which prevented Lorentz from discovering special relativity. He believed that the time t of an observer at rest with respect to the aether was the true time, whereas the quantity t' (which Lorentz called local time) of another observer, moving with respect to the first, was merely an abstraction that did not represent anything in the world[9]. In 1915 in a note added to the second edition of his book *The Theory of Electrons and Its Applications to the Phenomena of Light and Radiant Heat* Lorentz himself admitted the failure of his approach [7]:

> The chief cause of my failure was my clinging to the idea that the variable t only can be considered as the true time and that my local time t' must be regarded as no more than an auxiliary mathematical quantity. In Einstein's theory, on the contrary, t' plays the same part as t; if we want to describe phenomena in terms of x', y', z', t' we must work with these variables exactly as we could do with x, y, z, t.

The second example is another case in the history of physics when an inadequate philosophical position prevented even great scientists from making a discovery. It was Poincaré who first noticed (about three years before Minkowski gave his talk *Space and Time* in September 1908 [6]) that the Lorentz transformations had a geometric interpretation as rotations in an *abstract* four-dimensional space with time as the fourth dimension [8, p. 168]. Unfortunately, even such deep thinker and brilliant scientist as Poincaré appeared to have seen nothing extraordinary in the idea of a four-dimensional mathematical space *used to describe physical phenomena*. The most probable explanation of why he did not develop further this revolutionary idea might have been his conventionalism. Poincaré believed that our physical theories are only convenient descriptions of the world and therefore one could not ask which the correct theory is; in his view it is merely a matter of *convenience* which theory we would use in a given situation[10]. Poincaré appeared to have

---

[9]This is a rather sad example since Lorentz reified an unsuccessful abstraction – the aether – whereas he did not regard t' as representing something real. However, later it turned out that t' was a revolutionary successful abstraction when Einstein realized that both t and t' represented real properties of the world.

[10]Poincaré's philosophical views on the nature of physical theories appeared to have also prevented him (like Lorentz) from discovering special relativity before Einstein

8

thought that it is our choice (since a theory in his view was just a description) whether we will describe the world with a three-dimensional mathematical space and one dimensional mathematical space (representing time) or with a four-dimensional pseudo-Euclidean mathematical space (representing space and time as an inseparable entity); then we cannot ask whether the world is three- or four-dimensional.

Poincaré's conventionalism is clearly evident in what he wrote on spacetime several months before his death [9]:

> Everything happens as if time were a fourth dimension of space, and as if four-dimensional space resulting from the combination of ordinary space and of time could rotate not only around an axis of ordinary space in such a way that time were not altered, but around any axis whatever ...

> What shall be our position in view of these new conceptions? Shall we be obliged to modify our conclusions? Certainly not; we had adopted a convention because it seemed convenient and we had said that nothing could constrain us to abandon it. Today some physicists want to adopt a new convention. It is not that they are constrained to do so; they consider this new convention more convenient; that is all. And those who are not of this opinion can legitimately retain the old one in order not to disturb their old habits. I believe, just between us, that this is what they shall do for a long time to come.[11]

By contrast, Minkowski acknowledged that "it is to the credit of A. Einstein who first realized clearly that ...t and t' should be treated equally" [6, p. 117] and made a huge step forward. He recognized that the equal status of the times t and t' of two observers in relative motion implies that having different times the observers must also have *different spaces* (since time is perpendicular to space), which is only possible in a four-dimensional world; *observers in relative motion cannot have*

---

since his views precluded him from comprehending the profound physical meaning of the principle of relativity.

[11]Poincaré even appeared to have thought that the spacetime convention would not be advantageous: "It quite seems, indeed, that it would be possible to translate our physics into the language of geometry of four dimensions. Attempting such a translation would be giving oneself a great deal of trouble for little profit, and I will content myself with mentioning Hertz's mechanics, in which something of the kind may be seen. Yet, it seems that the translation would always be less simple than the text, and that it would never lose the appearance of a translation, for the language of three dimensions seems the best suited to the description of our world, even though that description may be made, in case of necessity, in another idiom" [10].

different spaces in a three-dimensional world since there is just *one* (and therefore *absolute*) space in such a world. Minkowski noted that "The concept of space was shaken neither by Einstein nor by Lorentz" [6, p. 117] and stressed that [6, p. 114]:

> Hereafter we would then have in the world no more *the* space, but an infinite number of spaces analogously as there is an infinite number of planes in three-dimensional space. Three-dimensional geometry becomes a chapter in four-dimensional physics.

Minkowski appeared to have been fully aware that the four-dimensional mathematical space with time as the fourth dimension is not just a convenient mathematical abstraction, but reflects a real four-dimensional world, which he called the absolute world. In it observers in relative motion still can formally introduce their spaces and times but they are rather *fictitious* since the absolute world, or spacetime, is a monolithic entity that is not objectively divided into space and time. This explains why he started his paper "Space and Time" with the now famous introduction, which unequivocally announced the revolution in our views of space and time [6, p. 111]:

> The views of space and time which I want to present to you arose from the domain of experimental physics, and therein lies their strength. Their tendency is radical. From now onwards space by itself and time by itself will recede completely to become mere shadows and only a type of union of the two will still stand independently on its own.

In 1921 Eddington specifically addressed the question of the *reality* of Minkowski's four-dimensional world when he discussed the fact that not only do observers in relative motion have different times but they also have different spaces as Minkowski discovered[12] and stressed that it is not a mere mathematical construction [11, p. 803]:

---

[12] In his book *Space, Time and Gravitation* Eddington is much more explicit [12, p. 51]: "However successful the theory of a four-dimensional world may be, it is difficult to ignore a voice inside us which whispers "At the back of your mind, you know that a fourth dimension is all nonsense." I fancy that that voice must often have had a busy time in the past history of physics. What nonsense to say that this solid table on which I am writing is a collection of electrons moving with prodigious speeds in empty spaces, which relatively to electronic dimensions are as wide as the spaces between the planets in the solar system! What nonsense to say that the thin air is trying to crush my body with a load of 14 lbs. to the square inch! What nonsense that the star-cluster, which I see through the telescope obviously there now, is a glimpse into a past age 50,000 years ago! Let us not be beguiled by this voice. It is discredited."

It was shown by Minkowski that all these fictitious spaces and times can be united in a single continuum of four dimensions. The question is often raised whether this four-dimensional space-time is real, or merely a mathematical construction; perhaps it is sufficient to reply that it can at any rate not be less real than the fictitious space and time which it supplants.

There is one more reason for which I have chosen the case of Poincaré's failure to realize the profound physical meaning of the four-dimensional mathematical space as one of the example of how incorrect views on the nature of physical theories can prevent even a scientist as great as Poincaré from making a significant contribution to physics – I find it disturbing that even now there are physicists who still appear to share similar views (with potentially negative consequences for the advancement of fundamental physics) as seen from a very recent example of another similar failure[13] to understand properly the nature of Minkowski's four-dimensional world [4, p. 9]:

What about spacetime itself? Is that real? Spacetime is a (3+1)-dimensional mathematical continuum. Even if you are a mathematical Platonist, I would urge you to consider that this continuum is nothing more than an extremely effective way to represent relations between distinct events ... So spacetime is an abstract four-dimensional mathematical continuum of points that approximately represent phenomena whose spatial and temporal extension we find it useful or necessary to ignore. The device of spacetime has been so powerful that we often reify that abstract bookkeeping structure, saying that we inhabit a world that is such a four- (or, for some of us, ten-) dimensional continuum.

---

[13]Hegel would probably remind us of the relevance of his observation even to matters of science: "what experience and history teach is this – that peoples and governments never have learned anything from history, or acted on principles deduced from it" [13]. To those who do not care what a philosopher would say, I would suggest that they examine both Minkowski's own arguments [6] for the reality of the absolute four-dimensional world and the arguments (or rather statements) against that view. Or, assume that spacetime were indeed only a mathematical construction and see whether the kinematical relativistic effects would be possible [14].

1.2 The role of conceptual analyses

Another reason that might have played some role in the lack of revolutions in physics in the last several decades may be the decreased if not abandoned employment of conceptual analyses in the search for new physical theories. Many physicists appear to regard them either as old fashioned or belonging to philosophy which seems to be supported by any search to see whether such analyses have been used in papers on open questions in fundamental physics published in the last several decades.

As the history of physics unambiguously proves that conceptual analyses are physics at its best, I think the most convincing way to demonstrate that is through several examples of how such analyses were pivotal for making discoveries and arriving at important results.

- It would hardly be overstated that Galileo was the first who systematically and productively used conceptual analyses of real and thought experiments which led him to

    - the idea of inertial motion that disproved the almost twenty century old (at that time) Aristotelian dogma that a moving body needs a mover; Galileo conceptually analyzed a number of real and thought experiments to prove that a free moving body moves on its own and does not need a mover.

    - the (Galileo's) principle of relativity (crucially based on the idea of inertial motion) with whose help Galileo disproved the arguments against a moving Earth, which paved the way for the acceptance of the heliocentric planetary system.

    - the important discovery that all bodies (heavy and light) fall equally (with the same acceleration) towards the Earth's surface.

- Newton also significantly relied on conceptual analyses

    - when he formulated his second law and particularly his third law[14].

---

[14]Newton's third law – action and reaction – "is the only one of the laws that Newton himself did not assign to his illustrious predecessor, Galileo. Newton seems to have come upon this law while contemplating the varieties of elastic and inelastic collisions which are introduced in the scholium to the laws" [15, p. 117].

– when he formulated his law of *universal* gravitation[15].

- The power of conceptual analyses and why these analyses are indeed physics at its best is particularly manifested in the case of Einstein:

  – The persistent analysis of his thought experiment of racing a light beam[16] (which he first considered when he was sixteen) led him to the realization that time was not absolute, which was decisive for the discovery of special relativity. It is hardly necessary to stress the obvious – it were the conceptual analyses of real and thought experiments that allowed the young and inexperienced scientist (without exceptionally great background in mathematics) to formulate the special theory of relativity, whereas such scientific colossi as Lorentz and Poincaré failed. Einstein even believed that a crucial role for the discovery of special relativity played also his studies of philosophical ideas and especially those of David Hume, "whose treatise on understanding [*A Treatise of Human Nature*] I studied with fervor and admiration shortly before the discovery of the theory of relativity. It is very well possi-

---

[15]Conceptual analyses played a key role in extrapolating the force of gravity to the whole universe because "out in space, the existence of such a force of attraction is not manifested in relation to a directly experienced phenomenon, but is only inferred by logic and a theory of rational mechanics, on the grounds that the planets do not move uniformly straight forward" [15, p. 113].

[16]This thought experiment became a paradox for Einstein when he studied Maxwell's equations at the Polytechnic Institute in Zurich. In Maxwell's theory the velocity of light is a universal constant ($c = (\mu_0\epsilon_0)^{-1/2}$) which meant for Einstein (due to his trust in "the truth of the Maxwell-Lorentz equations in electrodynamics" and that they "should hold also in the moving frame of reference" [18, p. 139]) that if he travelled almost at the speed of light (relative, say, to Earth), a beam of light would still move away from him at velocity c, which is in Einstein's own words "in conflict with the rule of addition of velocities we knew of well in mechanics" [18, p. 139]. Later Einstein acknowledged that "the germ of the special relativity theory was already present in that paradox" [19, p. 166] and explained that his "solution was really for the very concept of time, that is, that time is not absolutely defined but there is an inseparable connection between time and the signal velocity. With this connection, the foregoing extraordinary difficulty could be thoroughly solved. Five weeks after my recognition of this, the present theory of special relativity was completed" [18, p. 139]. The paradox's vital role explains why he owed "more to Maxwell than to anyone" else [20, p. 152] and why the Michelson-Morley experiment had probably played no role in Einstein's thinking (interestingly, Minkowski had regarded the same experiment as providing the experimental support for his spacetime formulation of special relativity: "The views of space and time which I want to present to you arose from the domain of experimental physics, and therein lies their strength." [6]).

ble that without these philosophical studies I would not have arrived at the solution" [20, p. 61].

- A careful analysis of the experiments on the photoelectric effect led Einstein to the correct theoretical description where he appeared to have trusted Planck's idea of quantization of electromagnetic radiation more than Planck himself.

- Einstein's analyses of his famous thought experiments involving elevators (accelerating or at rest in a gravitational field) allowed him to formulate his equivalence principle, which led him to the revolutionary view of gravity as spacetime curvature. In 1915 all people on Earth knew gravity was a force; only a single person disagreed and insisted gravity was not a force, but a mere manifestation of the non-Euclidean geometry of spacetime. Fortunately, unlike sometimes in society, in science democracy is never allowed to overrule objectivity (since natural laws are not determined by voting).

• Two examples of how the employment of conceptual analyses and the right understanding of the nature of physical theories allowed Dirac to arrive at two important results:

- Dirac's brilliant conceptual analysis of interference of photons helped him to realize that every photon of a photon beam should participate in both components of the split beam and should *interfere only with itself* [17]

- Dirac shared the view that mathematical entities should in general have counterparts in the world, which helped him to predict the existence of anti-particles.

• Despite that in their 1935 paper [17] Einstein, Podolsky and Rosen failed to prove that quantum mechanics was incomplete their inge-

---

[17] This analysis is perhaps one of the most efficient – such a significant feature of the quantum objects is proved in just several sentences [16]: "Suppose we have a beam of light consisting of a large number of photons split up into two components of equal intensity. On the assumption that the intensity of a beam is connected with the probable number of photons in it, we should have half the total number of photons going into each component. If the two components are now made to interfere, we should require a photon in one component to be able to interfere with one in the other. Sometimes these two photons would have to annihilate one another and other times they would have to produce four photons. This would contradict the conservation of energy. The new theory, which connects the wave function with probabilities for one photon, gets over the difficulty by making each photon go partly into each of the two components. Each photon then interferes only with itself. Interference between two different photons never occurs."

nious analysis revealed an essential feature of the quantum world – quantum entanglement.

1.3   Plan of the book

The research strategy outlined above (and particularly its two major components) will be employed in the analysis of inertia and gravitation. Prominent role will be given to conceptual analyses whose essence is the most valuable component in the birth of a scientific discovery, extracted from the analysis of breakthroughs and important results in physics – *exploring the internal logic of ideas*[18] (deducing all logical consequences of an idea and examining their implications through thought and real experiments).

Through these analyses I hope you will enjoy what appears to be "an eternal golden braid" [21] in all phenomena involving inertia and gravitation – the interplay of natural (non-resistant) and resistant motion. The best example is a falling body – Aristotle explained its motion as natural (not caused by anything), then Newton managed to convince the world that that motion is forced (and therefore resistant since a (gravitational) force is needed, according to Newton's second law, to overcome the resistance the falling body offers to its acceleration), and finally the circle (or rather the helix) was "completed" by Einstein who again had to convince the world to change their view on what causes the motion of a falling body – the body moves non-resistantly (by inertia or naturally) and no gravitational force is accelerating it; its acceleration is relative and results from the geodesic deviation of the geodesic worldlines of the body and the Earth's center (reflecting the fact that there are no parallel worldlines in the curved spacetime region around the Earth, which induces the spacetime curvature).

Chapter 2 starts with a brief account of how Aristotle divided motion in two categories – natural and violent – and continues with a discussion of how Galileo's analyses of real and thought experiments disproved Aristotle's view of motion and led him to the concept of inertial motion. Then it is discussed how again Galileo's real and thought experiments led him to the idea of inertia. Most of the chapter is devoted to an analysis of Newton's laws of motion, which demonstrates that they all are essentially based on the concept of inertia.

---

[18]The idea is to demonstrate that explicitly exploring the internal logic of fundamental ideas can provide shorter paths to scientific discoveries; such implicit analyses of the internal logic of profound (i) physical ideas (like those performed by Galileo, Newton, and Einstein), and (ii) mathematical ideas (like those carried out by Hilbert, Riemann, and especially Minkowski) already led to groundbreaking advancements.

Chapter 3 deals with Newton's theory of gravitation and the equivalence of inertial and gravitational mass. An examination of the internal logic of the Newtonian mechanics shows that there are serious problems with the concept of gravitational force, which could have been detected before the advent of general relativity.

Chapter 4 analyzes Newton's and Mach's views on the origin of inertia.

Chapter 5 explores the internal logic of Minkowski's spacetime formulation of special relativity with a special emphasis on whether Minkowski's view provides an insight into the nature of inertia.

Chapter 6 explores the internal logic of Einstein's profound idea of gravity as a manifestation of the non-Euclidean geometry of spacetime. The main raison is to determine whether according to general relativity gravitational phenomena are caused by gravitational interaction.

As Hermann Minkowski's groundbreaking paper "Space and Time" is extensively discussed in the book it is included as Appendix A.

Appendix B contains the Appendix by the Editor "On Relativistic Mass" published in: A. Einstein, *Relativity*, edited by V. Petkov (Minkowski Institute Press, Montreal 2018).

Appendices C and D include two of my papers for two reasons. First, they contain more detailed discussion of several specific issues. Second, the two papers provide a summary of the main new results in the book and will allow anyone who does not have time to read the whole book (we all have so much to read these days) to look only at the two papers. Appendix C contains the paper "On inertial forces, inertial energy and the origin of inertia" and Appendix D includes "Can Gravity be Quantized?"

References

[1] E.F. Taylor, J.A. Wheeler, *Spacetime Physics*, 2nd ed. (W.H. Freeman and Company, New York 1992), p 20

[2] K. Menger, Introduction to the 6th edition of E. Mach, *The Science of Mechanics* (La Salle, Illinois 1960) pp vii-viii

[3] D.C. Dennett, *Darwin's Dangerous Idea: Evolution and the Meanings of Life* (Simon and Schuster, New York 1996) p 21

[4] N. David Mermin, What's bad about this habit, *Physics Today*, May 2009 pp 8–9

[5] E. P. Wigner, The unreasonable effectiveness of mathematics in the natural sciences, *Communications on Pure and Applied Mathematics* **13** (1960) pp 1–14

[6] H. Minkowski, Space and Time. New translation in: H. Minkowski, *Space and Time: Minkowski's Papers on Relativity* (Minkowski Institute Press, Montreal 2012) pp 111–125 (http://minkowskiinstitute.org/mip/); included in this book as Appendix A

[7] H. A. Lorentz, *The Theory of Electrons and Its Applications to the Phenomena of Light and Radiant Heat*, 2nd ed. (Dover, Mineola, New York 2003) p 321

[8] H. Poincaré, "Sur la dynamique de l'électron", *Rendiconti del Circolo matematico Rendiconti del Circolo di Palermo* **21** (1906) pp 129–176

[9] H. Poincaré, *Mathematics and Science: Last Essays (Dernières Pensées)*, Translated by J.W. Bolduc (Dover, New York 1963) pp. 23–24

[10] H. Poincaré, *Science and Method*, In: *The Value of Science: Essential Writings of Henri Poincaré* (Modern Library, New York 2001) p. 438

[11] A.S. Eddington, The Relativity of Time, *Nature* **106**, pp 802-804 (17 February 1921); reprinted in: A. S. Eddington, *The Theory of Relativity and its Influence on Scientific Thought: Selected Works on the Implications of Relativity* (Minkowski Institute Press, Montreal 2015) pp. 27-30, p. 30

[12] A.S. Eddington: *Space, Time and Gravitation: An Outline of the General Relativity Theory* (Cambridge University Press, Cambridge 1920); new publication: A.S. Eddington, *Space, Time and Gravitation: An Outline of the General Relativity Theory*, with a Foreword by Dennis Dieks, (Minkowski Institute Press, Montreal 2017)

[13] G.W.F. Hegel, *The Philosophy of History*. In: *Great Books of the Western World*, Vol. 43, ed. by M.J. Adler (Encyclopedia Britannica, Chicago 1993) p 161

[14] V. Petkov, *Relativity and the Nature of Spacetime*, 2nd ed. (Springer, Heidelberg 2009) Chap 5

[15] I.B. Cohen, A Guide to Newton's Principia. In: Isaac Newton, *The Principia: Mathematical Principles of Natural Philosophy*, A new translation by I.B. Cohen, A. Whitman and J. Budenz (University of California Press, Berkeley 1999)

[16] P.A.M. Dirac, *Principles of quantum mechanics*, 4ed. (Oxford University Press, Oxford 1958) p 9

[17] A. Einstein, B. Podolsky, and N. Rosen, Can quantum-mechanical description of physical reality be considered complete? *Phys. Rev.* **47** (1935) p 777

[18] A. Pais, *Subtle Is the Lord: The Science and the Life of Albert Einstein* (Oxford University Press, Oxford 2005)

[19] A. Folsing, *Albert Einstein: A Biography* (Penguin Books, New York 1997)

[20] D. Brian, *Einstein: A Life* (Wiley, New York 1996)

[21] D.R. Hofstadter, *Gödel, Escher, Bach: An Eternal Golden Braid* (Basic Books, New York 1999)

2. The concept of inertia introduced by Galileo and developed by Newton

Despite that at first sight motion seems self-evident and self-explanatory, since ancient times thinkers have tried to understand its nature. In ancient Greece Aristotle (384 BC – 322 BC) devoted a significant part of his *Physics* to the analysis of moving bodies and concluded that in "some cases their motion is natural, in others violent and unnatural" [1, p. 339 (Book VIII)], that is, "all movement is either compulsory or according to nature" [1, p. 294 (Book IV)]. Natural motion or motion "according to nature" was mostly[1] defined on the basis of his view that everything in the terrestrial world consisted of four elements (unlike the heavens made of quintessence, i.e. a fifth element) – air, earth, fire, and water – and the assumption that the elements tend to move towards their natural places, e.g. "a body has a natural locomotion towards the center if it is heavy, and upwards if it is light" [1, p. 284 (Book III)]. In the case of unnatural (violent) motion "Everything that is in motion must be moved by something" [1, p. 326 (Book VII)].

As Aristotle's view of motion plays a central role in what we now call the Aristotelian physics (since his geocentric system of the world is based on that view), let me summarize it. Natural motion is harmonious, not forced motion. A body which moves naturally, moves on its own; it is not compelled to move by something external. For example a body falling towards the surface of the Earth moves naturally, since the basic element comprising the body – earth – tends to occupy its natural place

---

[1]Aristotle also defined a bit different type of natural motion: "Thus in things that derive their motion from themselves, e.g. all animals, the motion is natural (for when an animal is in motion its motion is derived from itself): and whenever the source of the motion of a thing is in the thing itself we say that the motion of that thing is natural" [1, p. 339 (Book VIII)].

– the center of the Earth. Natural motions on Earth (excluding animals) are downwards and upwards (air tends to go up), whereas the natural motion of celestial bodies is in circles. Unnatural (violent) motion is disharmonious since it is a forced motion. That is why, any object which moves (not vertically) must be moved by something.

Aristotle's view of motion and his geocentric model of the Universe with the Earth at the center (further developed by Ptolemy (90 – ca. 168)) dominated the pre-scientific understanding of the world for twenty centuries. The geocentric, or Ptolemaic, system had been universally accepted throughout those centuries despite that shortly after Aristotle another Greek – Aristarchos of Samos (310 BC – ca. 230 BC) – developed the first heliocentric model of our planetary system placing the Sun at the center. His model was in contrast to another alternative to the geocentric model, which was proposed before Aristotle by the Pythagorean Philolaus (ca. 470 BC – ca. 385 BC), in which all celestial bodies, including the Sun, revolved around a central fire; for more details of Aristarchos' model see [2].

Although there have been many attempts to explain why Aristarchos' heliocentric model of the solar system did not survive, I think the most plausible reason is that it contradicted Aristotle's view of motion. Here are some of the early examples of that contradiction summarized by Ptolemy in his *The Almagest* in which he implicitly used Aristotle's view of motion to show that if the Earth were not stationary but rotated around its axis[2] we would observe the effects of that motion (but such effects were never detected) [3]:

> They would have to admit that the earth's turning is the swiftest of absolutely all the movements about it because of its making so great a revolution in a short time, so that all those things that were not at rest on the earth would seem to have a movement contrary to it, and never would a cloud be seen to move toward the east nor anything else that flew or was thrown into the air. For the earth would always outstrip them in its eastward motion, so that all other bodies would seem to be left behind and to move towards the west.

These arguments against the view of a moving Earth remained unrefuted until the seventeenth century when Galileo (1564 – 1642) succeeded in refuting them by first refuting Aristotle's view of motion. Without the colossal work of Galileo it is extremely difficult to guess

---

[2]The view that the Earth orbits the Sun requires that it also turns about its axis to explain the change of day and night.

how many years would have been necessary for the acceptance of the second and more detailed heliocentric model of the solar system, proposed by Copernicus (1473 – 1543). The reason is that Copernicus did not disprove the arguments against the view that the Earth is moving.

2.1 Inertial motion and the power of exploring the internal logic of ideas

In the second chapter (The Second Day) of his book *Dialogue Concerning the Two Chief World Systems – Ptolemaic and Copernican* Galileo concentrated mostly on a single experiment – the tower experiment – which summarized the arguments that appeared to prove the motionlessness of the Earth. The tower argument can be formulated in the following way: if the Earth were not stationary but were turning around its axis, then a stone dropped from the top of a tower (Fig. 2.1) would not fall at the base of the tower (point A), but at point B, since during the time taken by the stone to fall, the tower (being carried by the Earth's turning) would travel the distance AB. Galileo realized that since it was Aristotle's view of motion – everything that is in motion must be moved by something – which is behind this argument[3], it could not be true.

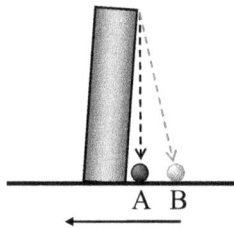

Figure 2.1: According to the supporters of the Ptolemaic system of the world, if the Earth were revolving around its axis and a stone is dropped from the top of, say, the Leaning Tower in Pisa it would hit the ground at B, not at A.

It is quite possible that Galileo himself dropped stones from the top

---

[3] According to the Aristotelian view when the stone is held on the top of the tower it has a mover and moves horizontally together with the tower (and is prevented from following its natural tendency to fall toward the center of the Earth). However, when the stone is released it only moves naturally (falls), but there is no mover to force it to move also horizontally and the stone is left behind the tower and hits the ground at point B.

of the mast of both a moving ship (or observed such experiments) and saw that the stone always fell at the foot of the must [4, pp. 144-145]:

> For anyone who does will find that the experiment shows exactly the opposite of what is written; that is, it will show that the stone always falls in the same place on the ship, whether the ship is standing still or moving with any speed you please. Therefore the same cause holding good on the earth as on the ship, nothing can be inferred about earth's motion or rest from the stone falling always perpendicularly to the foot of the tower.

Instead of stating clearly that he did perform or observe the experiments on a moving ship, Galileo chose to demonstrate the power of logic to arrive at the same result: "Without experiment, I am sure that the effect will happen as I tell you, because it must happen that way" [4, p. 145]. He did that by exploring the internal logic of the idea of motion of a firm ball on "a plane surface as smooth as a mirror and made of some hard material like steel" [4, p. 145] (Galileo analyzed several similar thought experiments based on real observations). Galileo considered three cases of the motion of a ball on a downward slope (Fig. 2.2 a), upward slope (Fig. 2.2 b), and on a horizontal plane (Fig. 2.2 c). In the first case the ball increases its velocity when released, whereas in the second case its velocity decreases after it was pushed upwards on the plane. For twenty centuries the supporters of Aristotle's view of motion (and, of course, Aristotle himself) have been well aware of plenty of instances of such trivial experiments, but did not bother to perform even a simple analysis of what they have been observing (which is inexplicable in the case of such a great thinker as Aristotle). Had Aristotle asked himself how a ball would move on a smooth horizontal surface (after being initially put in motion by a mover which is then removed), he undoubtedly would have arrived at the conclusion first realized by Galileo – the horizontal plane has neither downward nor upward slope, which means that the ball's velocity will neither increase nor decrease; therefore the ball will move with *constant* velocity. If the plane is infinite and the negligible friction is ignored the ball will move forever *on its own without a mover* (the friction only slows down and eventually stops the ball's eternal motion, but obviously has nothing to do with the newly discovered nature of motion).

The analyses of such experiments marked the end of Aristotle's view that a moving body requires a mover. According to the new view of motion, extracted through analyses from the observation of mundane physical phenomena, a body moves naturally (by inertia) when it moves uniformly (with constant velocity) on its own (without a mover), whereas forced (violent) motion occurs when a body is prevented from moving naturally.

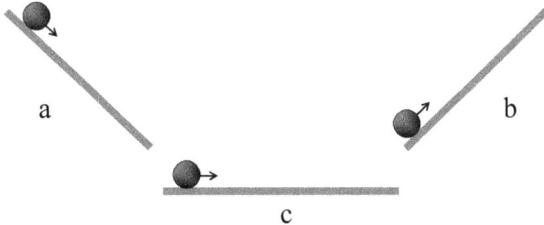

Figure 2.2: A simple analysis of the motion of a ball on inclined and horizontal planes led Galileo to the idea of inertial motion. Such an analysis could have been carried out by Aristotle and by all thinkers who for twenty centuries had supported his view of motion. Galileo did not possess any piece of extra knowledge that was not available to his predecessors.

Galileo's decision not to disclose whether he did or observed experiments of dropping a stone from the top of the mast of a moving ship might have been intended to convey a message to the future generations – whenever possible, try to explore the internal logic of ideas involved in open questions in order to arrive at resolutions even before performing real experiments. And he did give a convincing example – by employing the idea of inertial motion to the *thought* experiment of dropping a stone from the top of a tower or the mast of a moving ship he proved that the stone must fall at the foot of the tower or the mast, exactly at the place where it would fall if the earth or the ship were motionless: when the stone is released it preserves the horizontal component of its velocity (like the ball on the horizontal plane preserves its velocity after its mover is removed) and will fall at the base of the tower or the mast.

Therefore it does not matter whether or not a ship is at rest or moving; a stone dropped from the top of its mast will always fall at the base of the mast. No doubt this had been an impressive discovery at the time when being at rest and being in motion had been regarded as two distinct states. Galileo's analysis of thought (and perhaps real) experiments unequivocally demonstrated that such experiments involving inertial motion could not distinguish rest and motion.

By analyzing real and thought experiments Galileo refuted both Aristotle's view of motion and the arguments against the heliocentric model of the solar system, and arrived at two profound results:

- the idea of inertial motion – a free body moves on its own with constant velocity

- what we now call Galileo's principle of relativity – by performing mechanical experiments it is impossible to determine whether a body is at rest or moves with constant velocity.

I would complete the brief summary of these two great discoveries by Galileo with another brilliant example of his method of exploring the internal logic of ideas. The reason is to show that not only did he make contributions to physics, but he also gave us examples of doing physics successfully by showing how the basic element (conceptual analysis) of a productive research strategy works.

Galileo did not simply disprove Aristotle's view of motion (that a moving body should be *moved by something*) by demonstrating that free bodies move with constant velocity *on their own* (by inertia); he completely destroyed it as if as a revenge that for twenty centuries Aristotle's view unjustifiably held up the advancement of our understanding of the world. "Unjustifiably" because Aristotle himself and his followers during those centuries could have *realistically* discovered that that was a wrong view. You can judge whether this is really so from this example.

Aristotle appeared to have had a chance to disprove his own view when he had been struggling to explain the motion of projectiles. When a stone is thrown no mover is moving it. For some reason Aristotle gave a hasty explanation – the thrower's arm moves the air when the stone is thrown and the air acts as a mover of the stone. Galileo explored the internal logic of Aristotle's idea that projectiles are moved by the air. He started with a thought experiment in which a piece of cotton and a stone are placed on a table in an open field in a windy day. It is obvious that even not a strong wind will easily move the piece of cotton but not the stone. As the wind is motion of the air Galileo argued that the thought experiment unambiguously demonstrated that air moves light object more easily than heavy objects. This disproved Aristotle's explanation because if the air were the mover of projectiles than a thrown piece of cotton would go much farther than a thrown stone.

Galileo's analysis involved everyday phenomena, which had been repeatedly observed not only by Aristotle but also by ordinary people long before him. In other words, *Galileo did not use any piece of knowledge that was unavailable twenty centuries earlier*[4]. For this reason I think

---

[4]With this in mind, one may wonder whether the present situation in fundamental physics might be caused not by some still missing additional pieces of knowledge

Galileo's discoveries (inertial motion and Galileo's relativity principle) are delayed discoveries whose delay is difficult to comprehend and whose implications for the delayed advancement of our civilization are impossible to estimate. The only thing that could somehow attempt to explain how such a great thinker as Aristotle, who single-handedly created the science of logic, failed to see the problems with his view of motion, is that Galileo was the first who discovered and systematically used an innovative research strategy which turned out to be extraordinarily productive – conceptual analyses of real and thought experiments through exploring the internal logic of the ideas involved in those experiments.

2.2   The role of inertia in Newton's laws

Galileo's discoveries had profound implications for the development of physics, including modern physics. As we will now see Galileo's view of motion – inertial motion – is behind all Newton's laws. Galileo's relativity principle (along with Maxwell's electrodynamics as briefly explained in footnote 11 in the Introduction) led Einstein to the discovery of special relativity. In fact, applying Galileo's own method of exploring the internal logic of ideas to his relativity principle reveals that this principle is perhaps the greatest discovery of all time (without it, which led to special relativity, another great discovery – that gravity is spacetime curvature – would be impossible) – it *logically* contains the four-dimensional (Minkowski) formulation of special relativity[5] [6].

It was Galileo who arrived at the correct view of motion – that the natural motion of free bodies is motion with constant velocity of

---

that are necessary for a breakthrough, but by our failure to process the existing (and sufficient for the next revolution in physics) theoretical and experimental evidence. The major reason for the proposed research strategy (extracted from the successful approaches of the best physicists), outlined in the Introduction and used throughout the book, is to reduce the probability of future delayed discoveries in physics.

[5] In this sense special relativity is also a delayed discovery. Of course, it could not have been discovered in the seventeenth century, but in any case its discovery was realistically possible before the Michelson-Morley experiment. This is supported by the actual history of the discovery of special relativity – Einstein made it without taking account of this experiment (and probably without even knowing it); he analyzed only Galileo's relativity principle and Maxwell's electrodynamics (see footnote 11 in the Introduction and the references therein). A year before his death Einstein wrote "In my own development, Michelson's result has not had a considerable influence. I even do not remember if I knew of it at all when I wrote my first paper on the subject (1905). The explanation is that I was, for general reasons, firmly convinced that there does not exist absolute motion and my problem was only how this could be reconciled with our knowledge of electrodynamics. One can therefore understand why in my personal struggle Michelson's experiment played no role, or at least no decisive role" [5].

their own (by inertia) – but a rigorous theory of inertial motion and its implications for non-free bodies was created by Isaac Newton (1642 – 1727). Newton's first law of motion[6] is a statement of Galileo's law of inertia [7, p. 416]:

> Every body perseveres in its state of being at rest or of moving uniformly straight forward, except insofar as it is compelled to change its state by forces impressed.

Newton's first law describes the motion of a free particle that is not subject to any interactions. Indeed, it is the intrinsic feature of a particle to move *non-resistantly* by inertia when its motion is not disturbed by *any* influences that constitutes an objective criterion for a free particle. That is, *non-resistant* motion is a necessary and sufficient condition for a particle to be *free*. *A particle is subject to some interaction only if it resists its motion.*

Carefully examined, both Galileo's original idea of inertial motion and Newton's first law reveal two aspects of inertia – (i) a body moves by inertia on its own, and (ii) it resists when prevented from moving by inertia. First, a free body stays at rest or moves uniformly *on its own without a mover*, which means that it *offers no resistance* to its motion. "Every body perseveres in its state of being at rest or of moving uniformly" only because it does not resist its motion with constant velocity (including zero velocity). Second, a body "is compelled to change its

---

[6]It is interesting that Newton himself traced this law to ancient thinkers long before Galileo – back to Lucretius (ca. 99 BC – ca. 55 BC) and even to Aristotle (see [8]) and provided a quote from Aristotle's Physics (IV, 8) [1, p. 294]: "Further, no one could say why a thing once set in motion should stop anywhere; for why should it stop *here* rather than *here*? So that a thing will either be at rest or must be moved *ad infinitum*, unless something more powerful get in its way." In fact, however, Aristotle was against such a motion since he used that type of motion as an argument against the concept of void – if void existed a body would move in that unacceptable way (according to Aristotle). In regard to Lucretius, he did write in Book II of his *De Rerum Natura* (translated as *On the Nature of Things* or *The Way Things are*) that "all atoms are always moving... no atom ever rests coming through void" [9, p. 16], but this is not Galileo's and Newton's idea of inertia since according to Lucretius (who openly held Aristotle's view of motion [9, p. 29]) *only* individual atoms move in empty void in such a way. Ordinary bodies move differently, but "it does not work this way with single atoms which go along through empty void, unchecked by opposition... they are single units; they drive on, resistless, toward their first direction's impulse" [9, p. 17]. In [8] Cohen discussed Newton's and Descartes' views of inertia and especially Newton's use of the Latin expression *quantum in se est*, which "in its context provides a link between Newton's *Principia* and Descartes' *Principia* and in particular shows the origin of Newton's First Law of Motion in Descartes' First Law of Nature" [8, p. 147]. However, on whether Descartes had consistent understanding of inertia see [10]; he appeared to have regarded (elements of) circular motion as inertial as well.

state by forces impressed" because it *resists* any change in its inertial motion, and the forces are needed to *overcome* that resistance.[7] Newton finds the second aspect of inertia so important for understanding mechanical phenomena that he stated it in Definition 3 at the beginning of his *Principia* [7, p. 404]:

> *Inherent force of matter is the power of resisting by which every body, so far as it is able, perseveres in its state either of resting or of moving uniformly straight forward.*

Immediately after this definition Newton explains in more detail the resistance of matter when compelled to change its state, where he refers to the first aspect of inertia as inertia of matter [7, p. 404]:

> This force is always proportional to the body and does not differ in any way from the inertia of the mass except in the manner in which it is conceived. Because of the inertia of matter, every body is only with difficulty put out of its state either of resting or of moving. Consequently, inherent force may also be called by the very significant name of force of inertia. Moreover, a body exerts this force only during a change of its state, caused by another force impressed upon it, and this exercise of force is, depending on the viewpoint, both resistance and impetus: resistance insofar as the body, in order to maintain its state, strives against the impressed force, and impetus insofar as the same body, yielding only with difficulty to the force of a resisting obstacle, endeavors to change the state of that obstacle. Resistance is commonly attributed to resting bodies and impetus to moving bodies; but motion and rest, in the popular sense of the terms, are distinguished from each other only by point of view, and bodies commonly regarded as being at rest are not always truly at rest.

This quote is the best demonstration of Newton's crystal clear understanding of the reality of the force of inertia and of how inertia manifest itself – acting as both (depending on the view point) an active force (impetus) and a force of inertia (resistance). Our everyday experience

---

[7]The resistance of a body to a change in its uniform motion is easily deduced from Galileo's experiment with a ball moving on a horizontal plane – the fact that even a negligible friction does not immediately stop the ball implies that it tends to preserve its state of uniform motion and resists any force that prevents it from moving uniformly (by inertia).

provides us with plenty of examples of that duality – such an example is discussed in the beginning of the next section.

Newton's second and third law deal with the second aspect of inertia – the resistance a body offers when prevented from moving by inertia. The quantitative expression of that resistance is given in Newton's second law [7, p. 416]:

> A change in motion is proportional to the motive force impressed and takes place along the straight line in which that force is impressed.

Newton defined his famous equation $\mathbf{F} = m\mathbf{a}$ in the form

$$\mathbf{a} \propto \mathbf{F} \qquad \longrightarrow \qquad \mathbf{a} = \frac{1}{m}\mathbf{F}$$

which clearly reveals the physical meaning of mass in the coefficient of proportionality $1/m$ – as a greater force $\mathbf{F}$ is necessary to produce the same change in motion (i.e. the same acceleration $\mathbf{a}$) of a body of larger mass, the mass represents the *resistance* of a body to a change in its motion. For this reason *mass is defined as the measure of the resistance a body offers to its acceleration.*

Regarding Newton's second law, two things should be kept in mind:

- The profound meaning of this law is that *the force is only needed to overcome the resistance a body offers to its acceleration.* If a body offers no resistance to its motion, it is a free body (not subject to any force, i.e. interaction).

- Sometimes it is tempting to regard Newton's first law as a consequence of the second law – if $\mathbf{F} = 0$ then the first law, formally represented as $\mathbf{v} = const$, appears to follow: $\mathbf{a} \equiv d\mathbf{v}/dt = 0 \Rightarrow \mathbf{v} = const$. I think the best reaction to such a temptation (in addition to constantly remembering that physics is not mathematics) is to ask yourself why Newton needed a separate the first law and to recall how wisely he formulated his second law (so he clearly knew what he was doing). That if a particle is subject to no force its velocity is constant, simply means that Newton's two laws are consistent. The first law *defines* both inertial motion and a free particle by reflecting the *experimental* fact that a free particle moves *non-resistantly* on its own, i.e. that such a particle moves by inertia. Also, many physicists take the view that Newton's first law defines a special class of reference frames – inertial reference frames I – where Newton's laws hold.[8]

A body's reaction to a force that disturbs its inertial motion is captured in Newton's third law when applied to moving bodies [7, p. 417]:

> To any action there is always an opposite and equal reaction; in other words, the actions of two bodies upon each other are always equal and always opposite in direction.

Here is the explanation of this law in Newton's own words [7, p. 417]:

> Whatever presses or draws something else is pressed or drawn just as much by it. If anyone presses a stone with a finger, the finger is also pressed by the stone. If a horse draws a stone tied to a rope, the horse will (so to speak) also be drawn back equally toward the stone, for the rope, stretched out at both ends, will urge the horse toward the stone and the stone toward the horse by one and the same endeavor to go slack and will impede the forward motion of the one as much as it promotes the forward motion of the other. If some body impinging upon another body changes the motion of that body in any way by its own force, then, by the force of the other body (because of the equality of their mutual pressure), it also will in turn undergo the same change in its own motion in the opposite direction.

The last sentence demonstrates how the third law is applied to moving bodies. It is clearly seen that in such cases the third law reflects the fact that when a free particle is subject to an acting force $\mathbf{F}_a$ it offers an equal and opposite reaction $-\mathbf{F}_i$ by *resisting* the acting force, that is,

$$\mathbf{F}_a = -\mathbf{F}_i.$$

Therefore inertia is also behind Newton's third law when applied to moving physical objects as is further demonstrated in the example below.

---

[8]The first law defines an inertial reference frame in the following way – a reference frame I is inertial if free particles (subject to no forces) move with constant velocity relative to I. Sometimes a concern is expressed that such a definition involves a vicious circle [11] – a particle moves at constant velocity if it is not subject to any forces, but a criterion that no forces are acting on the particle is that it travels at constant velocity. Deep understanding of the nature of inertia and particularly of its two aspects makes it possible to see that there is no vicious circle – a rigorous criterion that no forces are acting on a particle is that it offers *no resistance* to its motion.

2.3 Are inertial forces real?

The most direct answer to this question is given by Newton's third law. Consider two balls A and B on a collision course, each of which moves at constant velocity relative to a stationary inertial reference frames (Fig. 2.3). Initially, at moment t_1 the balls are at some distance apart. At the next moment t_2 they collide.

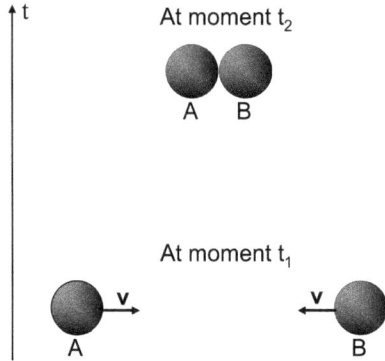

Figure 2.3: Two balls A and B move by inertia at moment t_1 and collide at moment t_2. Everything is symmetric in this example – A *resists* the acting force with which B prevents it from moving by inertia, which means that the reaction (i.e. the resistance) force with which A acts back on B is inertial in origin; from B's perspective it is A that prevents B from preserving its inertial motion and B exerts a reaction (i.e. inertial) force on A to counter the acting force coming from A. It is clearly seen that in cases like this *acting and reaction forces are both inertial forces.*

As both balls move by inertia when they collide each of them prevents the other from preserving its inertial motion. As a result each ball resists the change in its state of motion and exerts a resistance force on the other ball. It is evident that *both* resistance forces are inertial in origin and by Newton's third law are equal and opposite in direction. Not only does this example demonstrate that inertial forces are as real as the acting forces, but also that in situations like the one depicted in Fig. 2.3 the acting forces themselves are inertial.

The example in Fig. 2.3 shows that one has to be careful when distinguishing acting and inertial force. At first sight it seems perfectly reasonable to think that "inertial forces are not a cause, but a consequence, of accelerated motion" [12], but as we just saw the inertial forces with which balls A and B act on each other are both a cause and consequence – A's inertial force causes the inertial force with which B resists the change in its inertial motion and vice versa.

There should be no confusion concerning the reality of inertial forces when it is clearly stated that these forces are the *resistance* forces with which bodies moving by inertia oppose any change in their state of inertial motion. It is precisely the inertial forces that allow the detection of acceleration; as the acceleration of a body can be experimentally determined, acceleration is absolute or frame-independent.

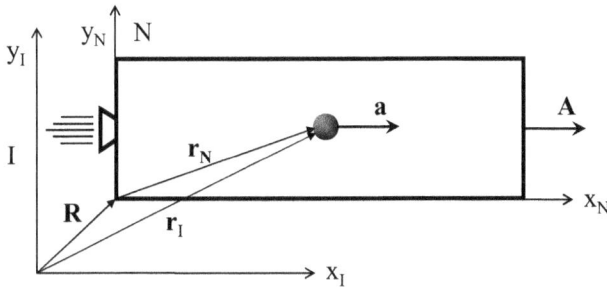

Figure 2.4: The accelerated motion of a ball is described in an accelerating elevator (a non-inertial reference frame N).

However, when we use non-inertial (accelerated) reference frames and want to continue using Newton's laws in such reference frames, we have to regard some force-like quantities as fictitious inertial forces. Consider a ball of mass m in a non-inertial (accelerating) reference frame N (e.g. an accelerating elevator) as shown in Fig. 2.4. If the ball is accelerating we can write Newton's second law in an inertial reference frame I where it holds:

$$m\mathbf{a} = \mathbf{F},$$

where $\mathbf{a} = \ddot{\mathbf{r}}_I$ is the ball's absolute acceleration, $\ddot{\mathbf{r}}_I$ is the second derivative with respect to time of $\mathbf{r}_I$, which is the instantaneous position of the ball relative to I, and $\mathbf{F}$ is the force accelerating the ball. If $\mathbf{r}_N$ is the instantaneous position of the ball with respect to N and $\mathbf{R}$ is the instantaneous position of N's origin relative to I we can write:

$$\mathbf{r}_I = \mathbf{R} + \mathbf{r}_N$$

Differentiating with respect to time twice (time is absolute in Newtonian mechanics[9]) and taking into account that $\ddot{\mathbf{R}} = \mathbf{A}$ is the acceleration of N (the elevator) relative to I we have:

$$\ddot{\mathbf{r}}_I = \mathbf{A} + \ddot{\mathbf{r}}_N.$$

Multiplying by m, rearranging, and replacing $m\ddot{\mathbf{r}}_I = m\mathbf{a}$ by $\mathbf{F}$ we arrive at an equation in N, which is similar to Newton's second law:

$$m\ddot{\mathbf{r}}_N = \mathbf{F} - m\mathbf{A}. \tag{2.1}$$

In the general case $|\mathbf{A}| \neq |\mathbf{a}|$. Newton's second law can be *formally* used in N if the force-like term $m\mathbf{A}$ is regarded as a fictitious force. This is one of the simple examples in physics where special care should be taken to ensure that mathematics should not take over physics (if this happens, confusion about the reality of inertial forces immediately follows). First, the acceleration $\ddot{\mathbf{r}}_N$ is *apparent*; like velocity it is frame-dependent, and does not represent anything independent of our description based on reference frames. The reason is that $\ddot{\mathbf{r}}_N$ is only a *fraction* of the frame-independent (i.e. absolute) acceleration $\mathbf{a} = \ddot{\mathbf{r}}_I$ (acceleration is an invariant in classical mechanics), which is not surprising since $\ddot{\mathbf{r}}_N$ is caused a force $(\mathbf{F} - m\mathbf{A})$ part of which $(\mathbf{F})$ is real and the other part $(m\mathbf{A})$ – fictitious. Second, even a quick examination shows why the term $m\mathbf{A}$ only resembles a force – the acceleration of the elevator $\mathbf{A}$ is multiplied by the mass of the ball (obviously, a force is expressed in terms of the mass and acceleration of the *same* body).

To understand better why the term $m\mathbf{A}$ is not a real force consider the case when the ball does not accelerate in I, that is when $\mathbf{a} = 0$. Assume that N is initially at rest relative to I in a region far away from gravitating masses, which means that the ball will be floating in the elevator (i.e. in N) as shown in Fig. 2.5. Then N starts to accelerate, but nothing accelerates the ball and it stays at rest relative to I. However, observed from within N the ball falls toward the floor of the elevator and its accelerated motion should be caused by a mysterious force $-m\mathbf{A}$ (resembling a gravitational force of a uniform gravitational field). Such an observation might have played a role in Einstein's thinking when he arrived at his equivalence principle – the term $-m\mathbf{A}$ has the form of a gravitational force caused by an apparent uniform gravitational field $\mathbf{G} = -\mathbf{A}$; and indeed, an observer in the accelerating elevator will see that all free objects in it fall towards the floor with the *same* acceleration $\mathbf{G}$.

---

[9] As a result of regarding time as absolute the first differentiation gives the classical law of addition of velocities $\dot{\mathbf{r}}_I = \dot{\mathbf{R}} + \dot{\mathbf{r}}_N$.

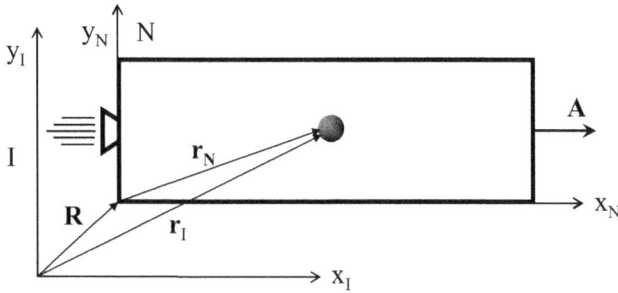

Figure 2.5: A ball moving by inertia is described in an accelerating elevator (a non-inertial reference frame N). As the ball does not accelerate (acceleration is absolute or frame-independent since the resistance an accelerated body offers to its acceleration is an experimental fact), it only appears to accelerate in N. To use Newton's second law a fictitious force should be introduced to account for the apparent acceleration.

In order to ensure that such cases and similar situations in physics do not cause confusion the following approach should be made part of a successful research strategy. Whenever dealing with a given situation, even before writing any equations the physical picture should be understood as much as possible. One should try to figure out *what is really happening regardless of how it is described in different reference frames.* Special care should be given to this task because in some cases such questions may be meaningless. But identifying such cases and trying to understand *why* they are meaningless is part of arriving at genuine understanding.

In our case what is absolute (frame-independent) is the acceleration of the ball. Unlike velocity, as explained above acceleration is absolute because it is experimentally detectable due to the resistance the ball (in this case) offers to its acceleration (as we will see in Chap. 5 special relativity provides a deeper insight into the absoluteness of acceleration and into the question of why it is measurable). So when N starts to accelerate it is clear that *what really happens* is that it is the elevator which accelerates and therefore it is its floor that approaches the ball (not the other way around). In N the ball only appears to fall toward the floor since no force acts on the ball. Then $\mathbf{F}=\mathbf{0}$ in Eq. (2.1) (which is also seen from the fact that the ball is at rest relative to I, that is, $\mathbf{r}_I$=const and it follows from $m\ddot{\mathbf{r}}_I = \mathbf{F}$ that $\mathbf{F} = 0$). By formally applying Newton's second law to describe the apparent acceleration $\ddot{\mathbf{r}}_N$

of the ball in N we have from Eq. (2.1)

$$m\ddot{\mathbf{r}}_N = -m\mathbf{A}. \qquad (2.2)$$

Formally, everything in Eq. (2.2) looks fine – the ball accelerates in N (in a direction opposite to $\mathbf{A}$) and its acceleration is caused by the force $\mathbf{F}_f = m\mathbf{A}$. However, taking into account the actual physical situation shows that the ball is at rest relative to I (and moves by inertia with respect to other inertial reference frames), which means that it does not resist its state of motion (which is experimentally verifiable by attaching an accelerometer to the ball, for example). That is why the ball's acceleration $\ddot{\mathbf{r}}_N$ is apparent and the force $\mathbf{F}_f$ is not real, but apparent, or fictitious.

The fictitious force $\mathbf{F}_f$ suddenly becomes real when the ball reaches (as observed in N) the floor of the elevator (Fig. 2.6). Proper understanding of the physical picture shows that there is nothing unusual – it is the floor that reaches the ball and starts to accelerate it; the ball resists the change in its inertial state and exerts a real inertial force back on the floor.

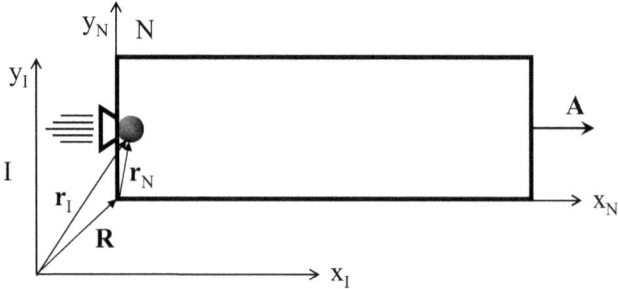

Figure 2.6: When the floor of the accelerating elevator reaches the ball, the ball starts to resist its acceleration by exerting a real inertial force on the floor.

Then as $\ddot{\mathbf{r}}_N = 0$ in N, Eq. (2.1), describing the motion of the ball in N, reduces to the equation

$$\mathbf{F} = m\mathbf{A},$$

since $\mathbf{F} = m\ddot{\mathbf{r}}_I$, where $\ddot{\mathbf{r}}_I = \ddot{\mathbf{R}}$ (since $\ddot{\mathbf{r}}_N = 0$ due to $\mathbf{r}_N = const$) is the acceleration of the ball in I, and $\ddot{\mathbf{r}}_I = \mathbf{A}$ (the ball's acceleration in I in this case is obviously equal to that of the elevator $\mathbf{A}$ in I). Also, the expression $m\mathbf{A}$ now makes sense – it is the product of the ball's mass and its acceleration in I.

To summarize – when observed from a non-inertial (accelerated) reference frame, a particle moving by inertia appears to accelerate in a direction opposite to the frame's acceleration (as shown in Fig. 2.5) and the particle's apparent acceleration can be formally regarded as caused by a fictitious inertial force; however, when the particle is prevented from moving by inertia and made to move with the frame's acceleration (as shown in Fig. 2.6), the particle resists its acceleration through a real inertial force.

The interplay between fictitious and real inertial forces is the same in all cases of non-inertial reference frames – translational (as the one discussed above), centrifugal, and Coriolis inertial forces are fictitious when a *free* particle (moving non-resistantly by inertia) *only appears to be accelerated by fictitious inertial forces* when described in non-inertial reference frames; when the particle is compelled to move with the non-inertial frames' accelerations, it starts to resist the change in its inertial motion and exerts *real inertial forces* on the mover that accelerates it.

The reality of inertial forces has an important consequence – like any force inertial forces also do work. This can be demonstrated with the following example. Imagine again two balls moving with constant velocity, i.e. by inertia, which collide as shown in Fig. 2.3. The standard explanation of the cause of the deformations the balls suffer is that it is the kinetic energy of the balls that is transformed into deformation energy. However, it is clear, when the reality of inertial forces is explicitly taken into account, that it is the *inertial energy* of the balls that is converted into deformation energy – each ball resist the change in its inertial motion and exerts a real inertial force on the other ball which does work on it; therefore the true physical meaning of kinetic[10] energy is inertial energy and not surprisingly the two energies are equally expressed by $mv^2/2$ as shown in Appendix C.

A final note regarding d'Alembert's principle due to its relevance to the question of the reality of inertial forces. In 1743 d'Alembert (1717 – 1783) published his *Traité de dynamique* where he formulated what is now called d'Alembert's principle. By treating inertial forces on the same footing as acting forces he managed to deal with situations in dynamics in the same way as with situations in statics[11]. D'Alembert

---

[10]The etymology of the word 'kinetic' already clearly indicates that kinetic energy is the energy of motion, but that motion with constant velocity is inertial and for this reason it is more helpful for the genuine understanding of physical phenomena to call the energy of motion by its proper name 'inertial energy.'

[11]D'Alembert's approach to view acting and inertial forces equally made it possible

suggested that Newton's second law $\mathbf{F} = m\mathbf{a}$ can be regarded as representing a dynamic equilibrium:

$$\mathbf{F} - m\mathbf{a} = 0, \tag{2.3}$$

where $-m\mathbf{a}$ is treated as an inertial force (as real as the acting force $\mathbf{F}$). Eq. (2.3) demonstrates even better than Newton's second law that the acting force $\mathbf{F}$ *neutralizes* the resistance (inertial) force $-m\mathbf{a}$ offered by the accelerated body of mass m in order to accelerate it. In other words, the forces $\mathbf{F}$ and $-m\mathbf{a}$ balance each other; hence dynamic equilibrium. The profound meaning of the d'Alembert's principle is fully revealed in Chap. 5 where we shall see that there is no difference between dynamic and static equilibrium in Minkowski spacetime.

References

[1] Aristotle: *Physics*. In: *Great Books of the Western World*, Vol. 7, ed. by M.J. Adler (Encyclopedia Britannica, Chicago 1993)

[2] C.M. Linton, *From Eudoxus to Einstein: A History of Mathematical Astronomy* (Cambridge University Press, Cambridge 2004) pp 38-45

[3] C. Ptolemy: *The Almagest*. In: *Great Books of the Western World*, Vol. 15, ed. by M.J. Adler (Encyclopedia Britannica, Chicago 1993) p 12

[4] G. Galileo: *Dialogue Concerning the Two Chief World Systems – Ptolemaic and Copernican*, 2nd edn. (University of California Press, Berkeley 1967)

[5] A. Pais, *Subtle Is the Lord: The Science and the Life of Albert Einstein* (Oxford University Press, Oxford 2005) p 172

[6] V. Petkov, *Relativity and the Nature of Spacetime*, 2nd ed. (Springer, Heidelberg 2009) Chap 3

[7] Isaac Newton, *The Principia: Mathematical Principles of Natural Philosophy*, A new translation by I.B. Cohen, A. Whitman and J. Budenz (University of California Press, Berkeley 1999)

[8] I. B. Cohen, *'Quantum in Se Est': Newton's Concept of Inertia in Relation to Descartes and Lucretius*, Notes and Records of the Royal Society of London, **19** (1964) pp 131-155

to apply the principle of virtual work not only in statics but in dynamics as well.

[9] Lucretius, *The Way Things are*. In: *Great Books of the Western World*, Vol. 11, ed. by M.J. Adler (Encyclopedia Britannica, Chicago 1993)

[10] René Descartes, *The World and Other Writings*. Translated and edited by S. Gaukroger (Cambridge University Press, Cambridge 1998), p xix

[11] R, Ferraro *Einstein's space-time: an introduction to special and general relativity* (Springer, Heidelberg 2007) p 9

[12] W. Benenson et al., *Handbook of Physics* (Springer, Heidelberg 2002) p 44

38

3. Newton's theory of gravitation and the equivalence of inertial and gravitational mass

3.1   How Newton discovered the law of universal gravity

According to the legend Newton received his first insight into what might be the nature of gravitation while observing a falling apple. Whether or not that really happened is a separate question, but it is the observation of falling bodies and his second law that certainly gave him to think. By the second law any acceleration is caused by a force and therefore it does appear to follow that a falling body should be subject to a force since it accelerates. With this in mind, Newton probably quickly reached the conclusion that it is the same force responsible for the weight of matter that also makes matter fall to Earth. Then he gradually arrived at the idea of universal gravity.

Here is how Newton reached the conclusion that the force of gravity acts as a centripetal force[1] (directed inward) and compels a body to orbit the Earth (Fig. 3.1), which is another example of the power of the method of exploring the internal logic of ideas [1]:

> That by means of centripetal forces the planets may be retained in certain orbits, we may easily understand, if we consider the motions of projectiles... for a stone projected is by the pressure of its own weight forced out of the rectilinear path, which by the projection alone it should have pursued,

---

[1]In the *Principia* Newton explains: "Hitherto we have called "centripetal" that force by which celestial bodies are kept in their orbits. It is now established that this force is gravity" [9, p. 806].

and made to describe a curve line in the air; and through that crooked way is at last brought down to the ground; and the greater the velocity is with which it is projected, the farther it goes before it falls to the earth. We may therefore suppose the velocity to be so increased, that it would describe an arc of 1, 2, 5, 10, 100. 1000 miles before it arrived at the earth, till at last, exceeding the limits of the earth, it should pass quite by without touching it.

Let AFB represent the surface of the earth, C its centre, VD, VE, VF, the curve lines which a body would describe, if projected in an horizontal direction from the top of an high mountain successively with more and more velocity... and, because the celestial motions are scarcely retarded by the little or no resistance of the spaces in which they are performed, to keep up the parity of cases, let us suppose either that there is no air about the earth, or at least that it is endowed with little or no power of resisting; and for the same reason that the body projected with a less velocity describes the lesser arc VD, and with a greater velocity the greater arc VE. and, augmenting the velocity, it goes farther and farther to F and G, if the velocity was still more and more augmented, it would reach at last quite beyond the circumference of the earth, and return to the mountain from which it was projected...

But if we now imagine bodies to be projected in the directions of lines parallel to the horizon from greater heights, as of 5, 10, 100, 1000, or more miles, or rather as many semi-diameters of the earth, those bodies, accord ing to their different velocity, and the different force of gravity in different heights, will describe arcs either concentric with the earth, or variously eccentric, and go on revolving through the heavens in those trajectories, just as the planets do in their orbs.

After demonstrating that the force of gravity, which compels matter to fall to the Earth, can also account for the orbital motion of celestial bodies, Newton proceeded to the formulation of his law of gravity. His thought experiment led him to conclude that "Gravity exists in all bodies universally and is proportional to the quantity of matter in each" [9, p. 810]. Then he determined how the force of gravity depended on distance. He started by proving that the centripetal forces of bodies that describe different circles tend toward the centers of the circles and are proportional to the squares of the orbital velocities of the bodies v and

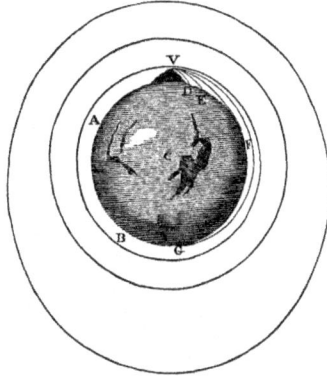

Figure 3.1: Newton's own figure [1, p. 513] depicting his thought experiment, which demonstrates "that by means of centripetal forces the planets may be retained in certain orbits."

inversely proportional to the radii of the circles r, that is, proportional to v^2/r: "the centripetal forces will be in a ratio compounded of the squared ratio of the velocities directly and the simple ration of the radii inverse;y" [9, p. 450]. Newton combined this result with what turned out to be also crucial for the discovery of the law of universal gravitation – Kepler's third law, which states that the square of the orbital period of a planet T is directly proportional to the cube of the radius of its orbit, that is, $T^2 \propto r^3$.

Then Newton showed that the gravitational force with which the Sun attracts the planets is inversely proportional to the square of their distances from the Sun. As the orbital velocity v of a planet in a circular orbit is equal to circumference of the orbit $2\pi r$ divided by the period T, the gravitational force is

$$F \propto \frac{v^2}{r} = \left(\frac{2\pi r}{T}\right)^2 \times \frac{1}{r} \propto \frac{r}{T^2}$$

and taking into account Kepler's third law $T^2 \propto r^3$

$$F \propto \frac{1}{r^2}.$$

Finally, taking into account that gravity "is proportional to the quantity of matter in each", the force of gravity between two bodies of masses M and m is:

$$F \propto \frac{Mm}{r^2}.$$

Or

$$F = G\frac{Mm}{r^2}, \qquad (3.1)$$

where G is what we now call the gravitational constant.

3.2   Newton and the mechanism of gravity

Although Newton had devoted more time than anyone else to the study of gravitational phenomena, he repeatedly insisted that his approach is different from the approaches of his predecessors and contemporaries. His dissatisfaction with them was caused by the fact that they were proposing different mechanisms of gravitation[2], which did not follow from the observed phenomena, but resulted from mixing the authors' metaphysical and physical views [1, p. 512]:

> The later philosophers pretend to account for it either by the action of certain vortices, as *Kepler* and *Des Cartes*; or by some other principle of impulse or attraction, as *Borelli, Honke,* and others of our nation; for, from the laws of motion, it is most certain that these effects must proceed from the action of some force or other.
>
> But our purpose is only to trace out the quantity and properties of this force from the phenomena, and to apply what we discover in some simple cases as principles, by which, in a mathematical way, we may estimate the effects thereof in more involved cases: for it would be endless and impossible to bring every particular to direct and immediate observation.
>
> We said, *in a mathematical way*, to avoid all questions about the nature or quality of this force, which we would not be understood to determine by any hypothesis; and therefore call it by the general name of a centripetal force, as it is a force which is directed towards some centre; and as it regards more particularly a body in that centre, we call it circum solar, circum-terrestrial, circum-jovial; and in like manner in respect of other central bodies.

---

[2]The search for a mechanism which can explain why bodies attract one another continued after Newton despite the triumph of his theory of gravitation [2]: "In the 18th and 19th centuries, countless attempts, now all but forgotten, were made to *explain* gravitation. Physicists hypostasized vortices, or tensions in media, or bombardments of the bodies by particles traversing space at random and driving, for instance, a stone toward the earth because the latter shields the stone against the particles coming from below."

In the General Scholium at the end of his *Principia* Newton gave even more explicit explanation why he did not propose any mechanism of gravity in his book [9, p. 943]:

> Thus far I have explained the phenomena of the heavens and of our sea by the force of gravity, but I have not yet assigned a cause to gravity. Indeed, this force arises from some cause that penetrates as far as the centers of the sun and planets without any diminution of its power to act, and that acts not in proportion to the quantity of the *surfaces* of the particles on which it acts (as mechanical causes are wont to do) but in proportion to the quantity of *solid* matter, and whose action is extended everywhere to immense distances, always decreasing as the squares of the distances. Gravity toward the sun is compounded of the gravities toward the individual particles of the sun, and at increasing distances from the sun decreases exactly as the squares of the distances as far out as the orbit of Saturn, as is manifest from the fact that the aphelia of the planets are at rest, and even as far as the farthest aphelia of the comets, provided that those aphelia are at rest. I have not as yet been able to deduce from phenomena the reason for these properties of gravity, and I do not feign hypotheses. For whatever is not deduced from the phenomena must be called a hypothesis; and hypotheses, whether metaphysical or physical, or based on occult qualities, or mechanical, have no place in experimental philosophy. In this experimental philosophy, propositions are deduced from the phenomena and are made general by induction. The impenetrability, mobility, and impetus of bodies, and the laws of motion and the law of gravity have been found by this method. And it is enough that gravity really exists and acts according to the laws that we have set forth and is sufficient to explain all the motions of the heavenly bodies and of our sea.

However, despite his famous pronouncement *hypotheses non fingo* (I do not feign hypotheses) in the above quotation Newton appeared to have never been satisfied with his inability to discover the true nature of gravity. What had been most troublesome to him is that gravitational attraction as described by his law constitutes an unacceptable action at a distance. Newton clearly expressed what he thought about such an action in his fourth letter to Richard Bentley of 25 February 1692 [4]:

That gravity should be innate, inherent, and essential to matter, so that one body may act upon another at a distance through a vacuum without the mediation of anything else, by and through which their action and force may be conveyed from one to another, is to me so great an absurdity, that I believe no man who has in philosophical matters a competent faculty of thinking can ever fall into it. Gravity must be caused by an agent acting constantly according to certain laws; but whether this agent be material or immaterial, I have left to the consideration of my readers.

3.3 Equivalence of inertial and gravitational masses

Newton struggled to define mass in the *Principia* as seen even from Definition 1 at its very opening: *"Quantity of matter is a measure of matter that arises from its density and volume jointly"* [9, p.403]. At the next page he explains: "I mean this quantity whenever I use the term "body" or "mass" in the following pages. It can always be known from a body's weight, for – by making very accurate experiments with pendulums – I have found it to be proportional to the weight."

Newton realized that the quantity of matter – which he called mass – had two aspects (manifestations) linked to the inertial and gravitating properties of matter. On the one hand, he was well aware that mass had something to do with the resistance of matter when accelerated: "because of the inertia of matter, every body is only with difficulty put out of its state either of resting or of moving" [9, p. 404]. On the other hand, Newton knew that mass also reflected the intrinsic feature of matter to attract gravitationally: *"Gravity exists in all bodies universally and is proportional to the quantity of matter in each"* [9, p. 810] and *"All bodies gravitate toward each of the planets, and at any given distance from the center of any one planet the weight of any body whatever toward that planet is proportional to the quantity of matter which the body contains"* [9, p. 806]. In Proposition 6 of the *Principia*'s Book 3 Newton described his experiments with pendulums which showed that mass is proportional to weight (the gravitational force), and which confirmed what had been known since Galileo[3] – that all bodies (heavy and

---

[3]Galileo found that when heavy and light bodies are dropped from the same height it takes them same time to reach the ground. There have been some doubts (probably prompted by Koyré [6]) that Galileo did such experiments but Galileo's texts themselves leave no such doubts. Here is an example which implies that he did

light) fall with the same acceleration [9, pp. 806-807]:

> Others have long since observed that the falling of all heavy bodies toward the earth (at least on making an adjustment for the inequality of the retardation that arises from the very slight resistance of the air) takes place in equal times, and it is possible to discern that equality of the times, to a very high degree of accuracy, by using pendulums. I have tested this with gold, silver,lead, glass, sand, common salt, water, and wheat... In these experiments, in bodies of the same weight, a difference of matter that would be even less than a thousandth part of the whole could have been clearly noticed.

I think, I. Bernard Cohen, the main translator of the modern translation of the *Principia* (1999), provided the best summary of Newton's understanding of mass [5]:

> He recognized that mass, the concept he had invented, is not only the measure of a body's resistance to a change of state or resistance to being accelerated, but is also the determinant of a body's reaction to a given gravitational field.

The manifestation of mass which reveals the inertial properties of matter is now called *inertial mass* (m_i) and is defined as the measure of a body's resistance to being accelerated; it is the coefficient of proportionality in Newton's second law of motion $\mathbf{F} = m_i \mathbf{a}$. The other manifestation of mass which reflects the gravitational properties of matter turned out to be more subtle. It itself has two manifestations – as the source of the gravitational field, and as a body's reaction to the gravitational force (in the same way as inertial mass is a body's reaction to an acting force which accelerates it). The first manifestation is called *active gravitational mass*, and the second – *passive gravitational mass*. What appears to be even more subtle is that the mass of a body acts

---

perform experiments with heavy and light bodies [7, pp. 447-448]: "Aristotle says that "an iron ball of one hundred pounds falling from a height of one hundred cubits reaches the ground before a one-pound ball has fallen a single cubit." I say that they arrive at the same time. You find, *on making the experiment* [italics added], that the larger outstrips the smaller by two finger-breadths, that is, when the larger has reached the ground, the other is short of it by two finger-breadths; now you would not hide behind these two fingers the ninety-nine cubits of Aristotle, nor would you mention my small error and at the same time pass over in silence his very large one."

as both active and passive gravitational mass. This is best seen from Newton's law of gravity (3.1):

$$F = G\frac{Mm}{r^2}.$$

Here each of the masses M and m of two interacting bodies acts as both active gravitational mass (creating the gravitational force that acts on the other mass) and passive gravitational mass (representing the reaction of each body to the gravitational force coming from the other body). If we write the law of gravity in the form of Newton's second law of motion (for the body of mass m which is subject to the gravitational force originating from mass M)

$$F = mg_M.$$

where the quantity

$$g_M = \frac{GM}{r^2}$$

has the dimensions of an acceleration (acceleration due to gravity), the mass m acts as passive gravitational mass, whereas the mass M plays the role of active gravitational mass. However we can also write (regarding the body of mass M as being subject to the gravitational force originating from mass m)

$$F = Mg_m.$$

where

$$g_m = \frac{Gm}{r^2}$$

and the roles of M and m are interchanged, which demonstrates that active and passive gravitational masses are clearly identical.

The equivalence of inertial and passive gravitational masses is usually demonstrated in the case of a falling body. Consider a body of some mass m which falls toward the surface of the Earth with an acceleration $a = g$. As this is accelerated motion we can describe it most generally in terms of Newton's second law (as caused by some force F):

$$F = m_i a. \tag{3.2}$$

In this case the body's mass m is regarded as its inertial mass m_i, which means that it is the measure of resistance the body offers to its acceleration. However, we can also describe the fall of the body by explicitly taking into account that it is the gravitational force that makes it accelerate downward:

$$F = m_g g. \tag{3.3}$$

In this case the body's mass m is treated as its passive gravitational mass m_g. As Newton's law of gravity is represented in the form of Newton's second law (3.2) we have no choice – the passive gravitational mass m_g should be also understood as the measure of the body's resistance to the acceleration due to gravity g. This understanding of the physical meaning of the passive gravitational mass m_g follows not only from the presentation of Newton's law of gravity in the form of his second law of motion, but also from the fact that the body is subject to the gravitational *force* originating from the Earth's mass M (according to the law of gravity) – it is the very physical meaning of the concept of force (reflected in Newton's third law[4]) which implies that a force is present only when it overcomes some resistance, and the measure of that resistance is the passive gravitational mass m_g. That a body which is subject to a gravitational force resists the change in its inertial motion is best seen in the case of the orbital motion of a planet – the gravitational force prevents the planet from moving by inertia along a straight line and it resists the change in its state.

So both the inertial mass m_i and the passive gravitational mass m_g turned out to be measures of the resistance a body offers to its acceleration. The passive gravitational mass of a body is the measure of its resistance to its acceleration caused *only* by the force of gravity, whereas the body's inertial mass is the measure of its resistance to its acceleration caused by *any* force (which means *including the force of gravity*). So even a brief conceptual analysis (by exploring the internal logic of the idea of mass as the measure of resistance to acceleration) indicates that the masses m_g and m_i are the same measure of resistance a body offers to its acceleration caused by whatever force.

However, the realization that m_i and m_g are equivalent had been achieved experimentally (therefore the experimental evidence confirmed the conclusion that *both* m_i and m_g are measures a body offers to its acceleration – a and g). Consider again a body falling toward the

---

[4]Usually this law is interpreted in a sense that the body acts back on the Earth from where the gravitational force originates. But this manifestation of Newton's third law reflects only the active gravitational masses of the body and the Earth – the Earth's active gravitational mass gives rise to a force which acts on the body, whose active gravitational mass acts back on Earth with its own gravitational force. But what about the body's passive gravitational mass? Especially given the fact that it is this mass that is the measure of the body's reaction to a gravitational force. Usually this question is avoided since regarding the passive gravitational mass as the measure of resistance to the gravitational acceleration posses a problem with Newton's third law – as gravity is not a contact force there is no object to which the reaction (resistance) of the body should be applied. But the fact is that if the body is subject to a force it should resist the force and such an argument could have been an early indication that Newton's concept of gravity as a force is problematic.

surface of the Earth. Its accelerated motion is described by Newton's second law (3.2). As the force in (3.2) in this case is the force of gravity (3.3) we have

$$m_g g = m_i a$$

or

$$a = \frac{m_g}{m_i} g, \tag{3.4}$$

which means that the observed acceleration a should be equal to the acceleration g caused by the force of gravity (originating from Earth's mass M) when $m_g = m_i$. In other words, a will be equal to g only if the resistance (represented by m_g) the falling body offers to its acceleration g caused by the gravitational force is equal to the body's resistance (m_i) to its *observed* acceleration a *regardless of the origin of the force responsible for it.*

The equivalence of the inertial and passive gravitational masses has been determined experimentally by proving that objects of different composition fall with the same acceleration[5]. Another piece of evidence demonstrating that $m_g = m_i$ is that all bodies dropped from a point at a distance r from the Earth's center fall with the *same observed acceleration* which is equal to

$$g = \frac{GM}{r^2}, \tag{3.5}$$

where, obviously, the Earth's mass M should be determined differently, not through measuring g. This indeed means that $m_g = m_i$ and by (3.4) the observed acceleration a of the falling bodies is equal to g. Therefore even Newton's law of gravity through Eq. (3.5) defines an acceleration g at a point r with which any body *regardless of its mass* will fall to Earth.[6]

---

[5]Galileo gave us another example of the power of the method of exploring the internal logic of ideas – he showed how Aristotle's idea that a heavy body falls faster than a light body leads to a contradiction [7, p. 446]: "If then we take two bodies whose natural speeds are different, it is clear that on uniting the two, the more rapid one will be partly retarded by the slower, and the slower will be somewhat hastened by the swifter... But if this is true, and if a large stone moves with a speed of, say, eight while a smaller moves with a speed of four, then when they are united, the system will move with a speed less than eight; but the two stones when tied together make a stone larger than that which before moved with a speed of eight. Hence the heavier body moves with less speed than the lighter; an effect which is contrary to your supposition. Thus you see how, from your assumption that the heavier body moves more rapidly than the lighter one, I infer that the heavier body moves more slowly."

[6]Now some authors prefer to call g defined by (3.5) the strength of the gravitational field or simply the gravitational field, but it is evident that g is the acceleration to which a body at a distance r from Earth's center is subjected.

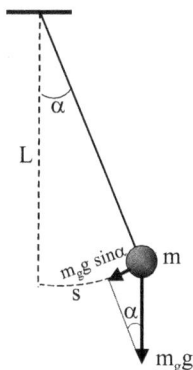

Figure 3.2: A simple pendulum.

Another method for testing that $m_g = m_i$ is the one mentioned in the beginning of this section – Newton himself tested the equivalence of the two masses with pendulums of different materials with 0.1 percent accuracy. Here is the idea of the simple pendulum experiments which played important role in our understanding of gravitation. A point mass m is suspended from a string with negligible mass which oscillates about the equilibrium (the lowest) position (Fig. 3.2). The tangential component $m_g g \sin \alpha$ of the gravitational force $F = m_g g$ (which contains the passive gravitational mass) is the force which causes the accelerated motion of the point mass along the arc s and Newton's second law in the tangential direction (whose left-hand side contains the inertial mass) is:

$$m_i \frac{d^2 s}{dt^2} = -m_g g \sin \alpha, \qquad (3.6)$$

where the minus sign reflects the fact that the tangential component of the gravitational force is always directed opposite to the motion of the point mass since it acts as a restoring force by preventing the point mass to continue to move *by inertia* (when moving *on its own* to the right or to the left *away* from the lowest position).

As $s = L\alpha$ and using the small angle approximation $\sin \alpha \approx \alpha$ (where α is measured in radians) we can write Eq. (3.6) in the form:

$$\frac{d^2 \alpha}{dt^2} = -\frac{m_g}{m_i} \frac{g}{L} \alpha. \qquad (3.7)$$

This equation describes the motion of the point mass for small amplitudes of oscillation and its mathematical form is the same as that of the equation describing simple harmonic motion

$$\frac{d^2x}{dt^2} = -\omega^2 x,\qquad (3.8)$$

where ω is the angular frequency. Comparing Eq. (3.7) and (3.8) allows us to write the angular frequency of the oscillations of the point mass of the simple pendulum:

$$\omega = \sqrt{\frac{m_g}{m_i}}\sqrt{\frac{g}{L}}.$$

Then the period of the motion of the simple pendulum is:

$$T = \frac{2\pi}{\omega} = 2\pi\sqrt{\frac{m_i}{m_g}}\sqrt{\frac{L}{g}}$$

and the periods of pendulums with different composition can be measured to determine whether they will depend on the composition of the point mass. Neither Newton no anyone else after him found such a dependence. Now students in introductory physics courses do experiments with pendulums and verify that their period is independent not only of the material of the point mass, but it is *independent of the mass itself*:

$$T = 2\pi\sqrt{\frac{L}{g}},$$

which means that $m_g = m_i$. This results is fully consistent with Eq. (3.5) and with the conclusion that in Newton's gravitational theory the trajectories in a gravitational field are particle (or mass) independent.

It is worth mentioning that even such simple experiments with pendulums allow us to test fundamental properties of matter as the equivalence of inertial and passive gravitational masses and to confirm conclusions achieved by conceptual analyses as those carried out by Galileo (like the one in footnote 4 above) and in this section.

3.4 Problems with Newton's view of gravity as a force

I will discuss briefly two problems with Newton's concept of gravitational force which could have been noticed before the advent of special relativity.

It turned out that only conceptual analyses can reveal the equivalence of inertial and passive gravitational mass in the case of falling bodies. However, the situation is different in the case of a body at rest in a gravitational field. The gravitational force acting on such a body is its weight which is expressed in the form of Newton's second law:

$$F = m_g g. \tag{3.9}$$

The question is how the passive gravitational mass should be understood in this case. Formally, as (3.9) is Newton's second law m_g should be also viewed as the measure of the body's resistance to its acceleration g. But obviously the body is not accelerating; it is not even moving.

A possible suggestion that Eq. (3.9) should not be treated as Newton's second law would not be convincing at all – in what sense it should not be Newton's second law; it has the form of that law. This will become clear in Chap. 6 when we will see that both the original Newton's second law $F = m_i a$ and $F = m_g g$ represent identical situations in spacetime – *deformations* (i) of the worldtube of a body by an acting force (in the case of $F = m_i a$) and (ii) of the worldtube of a body at rest on the Earth's surface (in the case of $F = m_g g$). As acceleration is absolute in spacetime since it is represented by the deformation of the worldtube of an accelerated body the situation represented in Eq. (3.9) finds a consistent but counter-intuitive explanation in relativity – the body is at rest on the Earth's surface but it does accelerate since its worldtube is deformed by the huge worldtube of the Earth.

The second problem is what in view of Newton's second law looks self-evident – that a falling body is subject to a gravitational force. In the previous section you saw how difficult it was to understand the physical meaning of passive gravitational mass for falling body and to distinguish it from the body's inertial mass. Had a second Galileo lived after Newton he might have explored the internal logic of Newton's view of gravity as a force and would have noticed the difficulties with m_g and m_i of a falling body. These difficulties would have given such a hypothetical thinker to think.

What the second Galileo would have explored is Galileo's results that all bodies fall with the same acceleration. Imagine that you drop some amount of sand from the leaning tower in Pisa. As all sand particles fall with the same acceleration the initial configuration of the dropped amount of sand will be preserved until it reaches the ground. This observation could have had unforeseen implications. Imagine that the sand particles were the constituents of a solid body – this will make it easier to realize that the real constituents of such a body falling in the Earth's gravitational field will stay in their equilibrium states. Therefore

52

the body will not be deformed, which means that it will not resist its fall, i.e. *it will not resist its accelerated motion.*

As a matter of fact, Galileo himself examined a similar thought experiment and virtually arrived at the conclusion that a falling body does not resist its fall [7, p. 447]:

> But if you tie the hemp to the stone and allow them to fall freely from some height, do you believe that the hemp will press down upon the stone and thus accelerate its motion or do you think the motion will be retarded by a partial upward pressure? One always feels the pressure upon his shoulders when he prevents the motion of a load resting upon him; but if one descends just as rapidly as the load would fall how can it gravitate or press upon him? Do you not see that this would be the same as trying to strike a man with a lance when he is running away from you with a speed which is equal to, or even greater, than that with which you are following him? You must therefore conclude that, during free and natural fall, the small stone does not press upon the larger and consequently does not increase its weight as it does when at rest.

However, if a body does not resist its acceleration it is not subject to any force since *a force is only needed to overcome a body's resistance to its acceleration.* At first, this would have been a very confusing conclusion since Newton's second law of motion makes it perfectly clear – every acceleration implies a force and vice versa. This conclusion would certainly bring memories of Aristotle who regarded the motion of falling bodies as natural or not forced. Then, a thorough analysis of the fact that all bodies fall with the same acceleration would have revealed an interesting similarity between free fall and inertial motion. According to Newton's first law of motion (and Galileo's own experiments [8] which had led him to the idea of inertial motion) different bodies move with the *same* velocity *by inertia* no matter whether they are heavy or light. So if heavy and light bodies fall with the same acceleration it is tempting the say that they move by inertia and because of this it does not matter whether they are heavy or light. However, the problem is obvious – how could they move by inertia if they accelerate?

References

[1] Isaac Newton, *The System of the World*. In: Isaac Newton, *Mathematical Principles of Natural Philosophy*, Translated by A. Motte (Daniel Adee, New York 1846) pp 512-513

[2] K. Menger, Introduction to the sixth edition of E. Mach, *The Science of Mechanics* (La Salle, Illinois 1960) pp vii-viii

[3] Isaac Newton, *The Principia: Mathematical Principles of Natural Philosophy*, A new translation by I.B. Cohen, A. Whitman and J. Budenz (University of California Press, Berkeley 1999)

[4] *Isaac Newton: Philosophical Writings*, A. Janiak (Ed.) (Cambridge University Press, Cambridge 2004) pp 102–103

[5] I.B. Cohen, A Guide to Newton's Principia. In: Isaac Newton, *The Principia: Mathematical Principles of Natural Philosophy*, A new translation by I.B. Cohen, A. Whitman and J. Budenz (University of California Press, Berkeley 1999) p 218

[6] Aleksandre Koyré, *Études galiléennes: La loi de la chute des corps. Descartes et Galilée.* (Hermann, Paris 1939)

[7] Galileo, *Dialogues Concerning Two Sciences.* In: S. Hawking (ed.), *On The Shoulders Of Giants*, (Running Press, Philadelphia 2002) pp. 399-626

[8] G. Galileo, *Dialogue Concerning the Two Chief World Systems – Ptolemaic and Copernican*, 2nd edn. (University of California Press, Berkeley 1967), Ch. 2 (The Second Day)

[9] I. Newton, *The Principia: Mathematical Principles of Natural Philosophy.* In: S. Hawking (ed.), *On The Shoulders Of Giants*, (Running Press, Philadelphia 2002) pp. 743-744

[10] Aristotle, *Physics* (Oxford University Press, New York 2008)

[11] R. Bentley, "Letter of Newton to Bentley, Trinity College, Jan. 17, 1692-3." In: *Works of Richard Bentley*, vol. 3 (London 1838) pp. 210–211. Quoted in: *Sir Isaac Newton's Mathematical Principles of Natural Philosophy and His System of the World* (Kessinger Publishing 2003) p. 634

54

4. Newton and Mach on the origin of inertia

Newton and Mach held strikingly different views on the origin of inertia and on the nature of motion and space. Newton believed that the very existence of inertia proved the absoluteness of motion and space, whereas Mach held the opposite view – that inertia was a result of relative motion.

4.1 Newton's view of inertia

As we saw in the previous chapter Newton regarded inertial forces as real forces of resistance which arise when physical bodies are prevented from moving by inertia. He did not attempt to explain their origin. But he seems to have suspected that the inertial forces had something to do with absolute space – that the resistance a body offers to its acceleration comes from the absolute space (as a result of the motion of the body through it). Indeed, as acceleration is absolute since it can be recognized by the inertial effects which are experimentally verifiable, it appears to follow that absolute space must exist because an accelerating body moves in space (an undeniable fact) and since accelerated motion is absolute the body accelerates with respect to an absolute space. For this reason, Newton believed that absolute (or true) motion existed and it could be discovered by its effects.

However, his understanding of the origin of inertia faced two major problems. First, if inertia were the resistance of the absolute space when bodies move through it, then why only accelerating bodies encountered that resistance, whereas bodies travelling with constant velocity (constant speed and constant direction) move through the absolute space freely. It is better to let Newton himself explain his views on this problem [1, p. 414]:

It is certainly very difficult to find out the true motions of individual bodies and actually to differentiate them from apparent motions, because the parts of that immovable space in which the bodies truly move make no impression on the senses. Nevertheless, the case is not utterly hopeless. For it is possible to draw evidence partly from apparent motions, which are the differences between the true motions, and partly from the forces that are the causes and effects of the true motions. For example, if two balls, at a given distance from each other with a cord connecting them, were revolving about a common center of gravity, the endeavor of the balls to recede from the axis of motion could be known from the tension of the cord, and thus the quantity of circular motion could be computed. Then, if any equal forces were simultaneously impressed upon the alternate faces of the balls to increase or decrease their circular motion, the increase or decrease of the motion could be known from the increased or decreased tension of the cord, and thus, finally, it could be discovered which faces of the balls the forces would have to be impressed upon for a maximum increase in the motion, that is, which were the posterior faces, or the ones that are in the rear in a circular motion. Further, once the faces that follow and the opposite faces that precede were known, the direction of the motion would be known. In this way both the quantity and the direction of this circular motion could be found in any immense vacuum, where nothing external and sensible existed with which the balls could be compared. Now if some distant bodies were set in that space and maintained given positions with respect to one another, as the fixed stars do in the regions of the heavens, it could not, of course, be known from the relative change of position of the balls among the bodies whether the motion was to be attributed to the bodies or to the balls. But if the cord was examined and its tension was discovered to be the very one which the motion of the balls required, it would be valid to conclude that the motion belonged to the balls and that the bodies were at rest, and then, finally, from the change of position of the balls among the bodies, to determine the direction of this motion. But in what follows, a fuller explanation will be given of how to determine true motions from their causes, effects, and apparent differences, and, conversely, of how to determine from motions, whether true or apparent, their causes and effects.

The second problem with Newton's understanding of the origin of inertia is Galileo's principle of relativity stating that absolute uniform motion cannot be discovered with mechanical experiments which implies that there is no absolute motion and therefore no absolute space. Here is Galileo's own demonstration of the nullity of all experiments designed to detect absolute motion [2, pp. 186–187]:

> For a final indication of the nullity of the experiments brought forth, this seems to me the place to show you a way to test them all very easily. Shut yourself up with some friend in the main cabin below decks on some large ship, and have with you there some flies, butterflies, and other small flying animals. Have a large bowl of water with some fish in it; hang up a bottle that empties drop by drop into a wide vessel beneath it. With the ship standing still, observe carefully how the little animals fly with equal speed to all sides of the cabin. The fish swim indifferently in all directions; the drops fall into the vessel beneath; and, in throwing something to your friend, you need throw it no more strongly in one direction than another, the distances being equal; jumping with your feet together, you pass equal spaces in every direction. When you have observed all these things carefully (though there is no doubt that when the ship is standing still everything must happen in this way), have the ship proceed with any speed you like, so long as the motion is uniform and not fluctuating this way and that. You will discover not the least change in all the effects named, nor could you tell from any of them whether the ship was moving or standing still. In jumping, you will pass on the floor the same spaces as before, nor will you make larger jumps toward the stern than toward the prow even though the ship is moving quite rapidly, despite the fact that during the time that you are in the air the floor under you will be going in a direction opposite to your jump. In throwing something to your companion, you will need no more force to get it to him whether he is in the direction of the bow or the stern, with yourself situated opposite. The droplets will fall as before into the vessel beneath without dropping toward the stern, although while the drops are in the air the ship runs many spans. The fish in their water will swim toward the front of their bowl with no more effort than toward the back, and will go

with equal ease to bait placed anywhere around the edges of the bowl. Finally the butterflies and flies will continue their flights indifferently toward every side, nor will it ever happen that they are concentrated toward the stern, as if tired out from keeping up with the course of the ship, from which they will have been separated during long intervals by keeping themselves in the air. And if smoke is made by burning some incense, it will be seen going up in the form of a little cloud, remaining still and moving no more toward one side than the other. The cause of all these correspondences of effects is the fact that the ship's motion is common to all the things contained in it, and to the air also.

Newton had been well-aware of Galileo's relativity principle. As Woodhouse explained [3, p. 8]:

In the manuscript *De motu corporum in mediis regulariter cedentibus* that Newton wrote two and half years before the laws of motion appeared in his *Philosophiae naturalis principia mathematica*, he had not three, but six laws of motion. The fourth was the principle of relativity.

The relative motion of bodies in a given space is the same whether the space is absolutely at rest or moves in a straight line without rotation.

Newton realized that the fourth law was a consequence of the first three (the three laws we know today); but he had other reasons for believing in an absolute standard of rest, which remains 'always similar and immovable'. In the *Principia*, the fourth law is reduced to the status of a corollary to the laws of motion.

That corollary to Newton's laws of motion – Corollary 5 – states: [1, p. 423]:

When bodies are enclosed in a given space, their motions in relation to one another are the same whether the space is at rest or whether it is moving uniformly straight forward without circular motion.

Then Newton continued by providing clear evidence that he was indeed fully aware of Galileo's principle of relativity and particularly with Galileo's experiments on a ship discussed in the text given above:

This is proved clearly by experience: on a ship, all the motions are the same with respect to one another whether the ship is at rest or is moving uniformly straight forward.

It seems Newton did not see any contradiction between Galileo's principle of relativity and his view that absolute (true) motions and absolute space (with respect to which the absolute motions occur) must exist. Indeed, the experiments Galileo discussed only fail to discover absolute *uniform* motion. The undeniable experimental fact that acceleration motion is detectable and therefore absolute might have been responsible for Newton's belief that relative motions are not true (absolute) motions since they are merely differences between absolute motions. That is why Newton had been searching for experiments to distinguish between absolute motion and relative motion. Such an experiment that he discussed at length is his famous bucket experiment [1, pp. 412-413]:

> The effects distinguishing absolute motion from relative motion are the forces of receding from the axis of circular motion. For in purely relative circular motion these forces are null, while in true and absolute circular motion they are larger or smaller in proportion to the quantity of motion. If a bucket is hanging from a very long cord and is continually turned around until the cord becomes twisted tight, and if the bucket is thereupon filled with water and is at rest along with the water and then, by some sudden force, is made to turn around in the opposite direction and, as the cord unwinds, perseveres for a while in this motion; then the surface of the water will at first be level, just as it was before the vessel began to move. But after the vessel, by the force gradually impressed upon the water, has caused the water also to begin revolving perceptibly, the water will gradually recede from the middle and rise up the sides of the vessel, assuming a concave shape (as experience has shown me), and, with an ever faster motion, will rise further and further until, when it completes its revolutions in the same times as the vessel, it is relatively at rest in the vessel. The rise of the water reveals its endeavor to recede from the axis of motion, and from such an endeavor one can find out and measure the true and absolute circular motion of the water, which here is the direct opposite of its relative motion. In the beginning, when the relative motion of the water in the vessel was greatest, that motion was not giving rise to any endeavor to recede from the axis; the water did not seek

the circumference by rising up the sides of the vessel but remained level, and therefore its true circular motion had not yet begun. But afterward, when the relative motion of the water decreased, its rise up the sides of the vessel revealed its endeavor to recede from the axis, and this endeavor showed the true circular motion of the water to be continually increasing and finally becoming greatest when the water was relatively at rest in the vessel. Therefore, that endeavor does not depend on the change of position of the water with respect to surrounding bodies, and thus true circular motion cannot be determined by means of such changes of position. The truly circular motion of each revolving body is unique, corresponding to a unique endeavor as its proper and sufficient effect, while relative motions are innumerable in accordance with their varied relations to external bodies and, like relations, are completely lacking in true effects except insofar as they participate in that true and unique motion. Thus, even in the system of those who hold that our heavens revolve below the heavens of the fixed stars and carry the planets around with them, the individual parts of the heavens, and the planets that are relatively at rest in the heavens to which they belong, are truly in motion. For they change their positions relative to one another (which is not the case with things that are truly at rest), and as they are carried around together with the heavens, they participate in the motions of the heavens and, being parts of revolving wholes, endeavor to recede from the axes of those wholes.

4.2   Mach's explanation of inertia

Newton's views of inertia, absolute motion and absolute space have been criticized by Berkeley [4] and particularly by Mach [5]. Both believed that it is meaningless to talk about absolute motion and absolute space since they are unobservable. For Mach [5, p. 280]:

> No one is competent to predicate things about absolute space and absolute motion; they are pure things of thought, pure mental constructs, that cannot be produced in experience.

Here is how Mach explained the difference between Newton's and his views on the nature motion and the appearance of inertial (centrifugal) forces [5, pp. 283-284]:

Let us now examine the point on which Newton, apparently with sound reasons, rests his distinction of absolute and relative motion. If the earth is affected with an *absolute* rotation about its axis, centrifugal forces are set up in the earth: it assumes an oblate form, the acceleration of gravity is diminishes at the equator, the plane of Foucault's pendulum rotates, and so on. All these phenomena disappear if the earth is at rest and the other heavenly bodies are affected with absolute motion round it, such that the same *relative* rotation is produced. This is, indeed, the case, if we start *ab initio* from the idea of absolute space. But if take our stand on the basis of facts, we shall find we have knowledge only of *relative* spaces and motions. *Relatively*, not considering the unknown and neglected medium of space, the motions of the universe are the same whether we adopt the Ptolemaic or the Copernican mode of view. Both views are, indeed, equally *correct*; only the latter is more simple and more *practical*. The universe is not *twice* given, with an earth at rest and an earth in motion; but only *once* with its *relative* motions, alone determinable. It is, accordingly, not permitted us to say how things would be if the earth did not rotate. We may interpret the one case that is given to us, in different ways. If, however, we so interpret it that we come into conflict with experience, our interpretation is simply wrong. The principles of mechanics can, indeed, be so conceived, that even for relative rotations centrifugal forces arise.

Newton's experiment with the rotating vessel of water simply informs us, that the relative rotation of the water with respect to the sides of the vessel produces *no* noticeable centrifugal forces, but such forces *are* produced by its relative rotation with respect to the mass of the earth and the other celestial bodies. No one is competent to say how the experiment would turn out if the sides of the vessel increased in thickness and mass till they were ultimately several leagues thick.

Mach stated Newton's view of absolute rotation – "By absolute rotation he understood a rotation relative to the fixed stars, and here centrifugal forces can always be found" [5, pp. 280-281] – and specifically criticized his bucket experiment [6, pp. 542-543]:

I cannot share this view. For me, only relative motions exist, and I can see, in this regard, no distinction between rotation and translation. When a body moves relatively to the fixed stars, centrifugal forces are produced; when it moves relatively to some different body, and not relatively to the fixed stars, no centrifugal forces are produced. I have no objection to calling the first rotation "absolute" rotation, if it be remembered that nothing is meant by such a designation except *relative rotation with respect to the fixed stars*. Can we fix Newton's bucket of water, rotate the fixed stars, and *then* prove the absence of centrifugal forces?

The experiment is impossible, the idea is meaningless, for the two cases are not, in sense-perception, distinguishable from each other. I accordingly regard these two cases as the *same* case and Newton's distinction as an illusion.

Mach appears to have believed that the inertial centrifugal forces arising in the water in Newton's bucket were caused by the distant stars, not by the absolute space as Newton believed. Such an explanation implies that it was the gravity of the distant stars that was responsible for inertia. However, such a straightforward at first sight attempt to reveal what Mach's view of the origin was of inertia was is seriously put into doubt by the following statement by Mach [6, p. 543]: "I never assumed that remote masses *only*, and not near ones, determine the velocity of a body; I simply spoke of an influence *independent* of distance." Obviously, if the gravity of all the masses in the universe were causing the centrifugal forces in the water in Newton's bucket, their influence would depend on the distance between them and the bucket.

4.3   Problems with Mach's relativism

The most definite and convincing rejection of Mach's relativism is provided by both special and general relativity as we will see in the next two chapters. Even Minkowski's four-dimensional formulation of special relativity is completely sufficient to demonstrate why Mach's relativism, like Poincaré's sterile conventionalism (discussed in the first chapter), is an unproductive general view of the world that superficially reflects the observed physical phenomena and prevents anyone who holds such a view from searching for deeper causes of phenomena by identifying the phenomena with with what is at their surface (and effectively declaring that there is nothing more to the phenomena than their surface immediately given in our senses).

It is well known that before 1905 Einstein{indexEinstein had been strongly influenced by Mach's relativism to such an extent that Einstein was "for general reasons, firmly convinced that there does not exist absolute motion" [7, p. 172]. Mostly due to Mach Einstein regarded all motion as relative. And indeed Einstein kept the term "relativity" in his general theory because he believed that in that theory acceleration should also be treated as relative. In his 1914 paper [8] Einstein repeated and extended Mach's argument for a relative acceleration. Here is how in 1921 Pauli explained the role of Mach's ideas in general relativity [9]:

> Mach ... had replaced absolute acceleration by acceleration relative to all the other masses of the universe. Einstein called this postulate 'Mach principle'. It has to be postulated, in particular, that the inertia of matter is solely determined by the surrounding masses. It must therefore vanish when all the other masses are removed, because it is meaningless, from a relativistic point of view, to talk of a resistance against absolute accelerations (relativity of inertia).

As we will see in the next two chapters acceleration is absolute in both special and general relativity and therefore there is no relativity of inertia. Indeed, in 1954 Einstein wrote regarding Mach: "As a matter of fact, one should no longer speak of Mach's principle at all"[1] [7, p. 288].

More specifically, we will see at the end of Chapter 5 why Mach's views on the nature of motion and inertia are wrong:

- Due to the fact that acceleration is absolute in relativity, rotation is absolute and therefore Mach's claim that the Ptolemaic and the Copernican system are equivalent is clearly erroneous.

- Contrary to Mach's claim, rotation and translation are clearly distinct motions in relativity.

- According to Mach's view if there were no other bodies in the universe, one could not talk about the state of the motion of single

---

[1]Occasionally, one can hear claims that Mach's ideas are behind the effect of frame-dragging in general relativity discovered by Lense and Thirring in 1918 [10]. Such claims, however, are based on misunderstanding since frame-dragging (due to the Earth's rotation, for example) is purely general relativistic effect which also demonstrates that rotation is absolute, whereas Mach held the opposite view – that it was relative. Not surprisingly, Mach's name was not mentioned in the 2011 paper "Gravity Probe B: Final Results of a Space Experiment to Test General Relativity" [11] which reported on the experimental confirmation of two predictions of general relativity, one of which is frame-dragging caused by the Earth's rotation.

particle. By contract, relativity says that due to the absoluteness of acceleration one can say whether or not a single particle in the universe accelerates (we will see this in the next chapter).

- According to Mach the origin of inertia is non-local since all the masses in the universe are responsible for the inertial forces, whereas relativity implies that inertia is a local phenomenon.

References

[1] Isaac Newton, *The Principia: Mathematical Principles of Natural Philosophy*, A new translation by I.B. Cohen, A. Whitman and J. Budenz (University of California Press, Berkeley 1999)

[2] G. Galileo: *Dialogue Concerning the Two Chief World Systems – Ptolemaic and Copernican*, 2nd edn. (University of California Press, Berkeley 1967)

[3] N.M.J. Woodhouse, *Special Relativity* (Springer, Heidelberg 2003)

[4] G. Berkeley, *Principles of Human Knowledge and Three Dialogues* (Oxford University Press, Oxford 1999) p. 75

[5] E. Mach, *The Science of Mechanics*, 6th ed. (La Salle, Illinois 1960)

[6] E. Mach, *The Science Of Mechanics*, 4th ed. (The Open Court Publishing Co., Chicago 1919)

[7] A. Pais, *Subtle Is the Lord: The Science and the Life of Albert Einstein* (Oxford University Press, Oxford 2005) p. 139

[8] A. Einstein, The Formal Foundation of the General Theory of Relativity, in: *The Collected Papers of Albert Einstein, Volume 6: The Berlin Years: Writings, 1914-1917* (Princeton University Press, Princeton 1997) p. 31.

[9] W. Pauli: *Theory of Relativity* (Dover, New York 1958) p. 179

[10] H. Thirring, Über die Wirkung rotierender ferner Massen in der Einsteinschen Gravitationstheorie, *Phys. Zeit.* **19** (1918) pp. 33-39; J. Lense and H. Thirring, Über den Einfluss der Eigenrotation der Zentralkörper auf die Bewegung der Planeten und Monde nach der Einsteinschen Gravitationstheorie, *Phys. Zeit.* **19** (1918) pp. 156-163

[11] C.W.F. Everitt et al, Gravity Probe B: Final Results of a Space Experiment to Test General Relativity, *Phys. Rev. Lett.* **106** (2011) p. 221101

5. Inertia and mass in Minkowski spacetime

In 1905 when Einstein published his paper on special relativity it appeared that the relativistic views of Berkeley and Mach had found their way in fundamental physics – even the name of Einstein's theory reflected those views. And indeed, Einstein declared that there is no absolute space (i.e. no ether) and therefore no absolute motion [1, p. 141]: "The introduction of a "light ether" will prove superfluous, inasmuch as in accordance with the concept to be developed here, no "space at absolute rest" endowed with special properties will be introduced, nor will a velocity vector be assigned to a point of empty space at which electromagnetic processes are taking place."

Einstein's rejection of the idea of absolute space and absolute motion, which helped him arrive at special relativity before Lorentz and Poincaré, was apparently based on two major reasons. First, as indicated in the previous chapter Einstein had been "for general reasons, firmly convinced that there does not exist absolute motion" [2, p. 172] mostly due to Mach's relativism. Second, Einstein had been impressed by the fact that the velocity of light is a fundamental constant $c = (\mu_0 \epsilon_0)^{-1/2}$ in Maxwell's electromagnetic theory since it is expressed through the fundamental (universal) constants μ_0 and ϵ_0 (the permeability and permittivity of vacuum or empty space). Lorentz, Poincaré and the other physicists at the turn of the nineteenth and the twentieth centuries seemed to have seen in the definition of c in terms of μ_0 and ϵ_0 only further support for the luminiferous ether since μ_0 and ϵ_0 were interpreted as describing the properties of this light carrying medium. For Einstein, however, the fact that c is a fundamental (*universal*) constant was extraordinary since *a fundamental constant should be a constant in all inertial reference frames.* But if this is so then c denotes the speed of light not in the ether (i.e. in the absolute space) but in *all* inertial reference frame. Whatever that meant for Einstein, it showed him that there was a serious problem with the concept of absolute space (if the absolute space existed, c should denote the speed of light with respect to it).

To make the above arguments even stronger Einstein began his 1905 paper "On the Electrodynamics of Moving Bodies," which contained his special theory of relativity, with an example intended to demonstrate that the observed phenomena do not seem to support the concept of absolute space and therefore of absolute motion and absolute rest [1, p. 140]:

> It is known that Maxwell's electrodynamics – as usually understood at the present time – when applied to moving bodies, leads to asymmetries which do not seem to attach to the phenomena. Let us recall, for example, the electrodynamic interaction between a magnet and a conductor. The observable phenomenon depends here only on the relative motion of conductor and the magnet ... Examples of a similar kind, and the failure of attempts to detect a motion of the earth relative to the "light medium," lead to the conjecture that not only in mechanics, but in electrodynamics as well, the phenomena do not have any properties corresponding to the concept of absolute rest, but that in all coordinate systems in which the mechanical equations are valid, also the same electrodynamic and optical laws are valid ...

The advent of special relativity led not only to the relativization of motion, but it also relativized simultaneity, time, length, and mass. At first sight it seemed that the triumph of relativism in physics was complete. However, the initial formulation of special relativity had the major hallmark of relativism – assuming that there is nothing more to phenomena (what we can observe) and ignoring the lessons from the history of science (especially the way Galileo discovered inertial motion and the principle of relativity) which demonstrate that the phenomena are representing only 'the surface of the world' since fundamental physical laws and profound features of the world manifest themselves through the phenomena.

Relativism is a sterile doctrine that is unable to decode the messages hidden in the phenomena and to lead to genuine and deep understanding of the world. Einstein was a happy exception. In his 1905 paper he used that doctrine only to reconcile Galileo's principle of relativity and Maxwell's electrodynamics and after that he gradually abandoned it. However, Einstein used his enormous intellectual potential to succeed in arriving at the theory of special relativity by using such an unproductive doctrine. It took him years and a huge intellectual effort to resolve the apparent paradox of racing a light beam, which he encountered at the age of sixteen as briefly discussed in the Introduction. When he finally succeeded to resolve the paradox, his genius helped him to arrive at a fundamental result (that time was not absolute) by using the unproductive relativistic doctrine because the "solution was really for the very concept of time, that is, that time is not absolutely defined but there is an inseparable connection between time and the signal velocity. With this connection, the foregoing extraordinary difficulty could be thoroughly solved. Five weeks after my recognition of this, the present theory of special relativity was completed" [2]. But even then the negative impact of that doctrine could be detected – Einstein had initially thought "that time is not absolutely defined but *there is an inseparable connection between time and the signal velocity*" (italics added), but later he realized (due to Minkowski's work) that the relativity of time had nothing to do with "the signal velocity" since that relativity was a manifestation of an *absolute* four-dimensional world (*only in such a world time can be relative*).

The major hallmark of relativism was clearly evident in the initial formulation of special relativity – a number of apparently paradoxical results were simply postulated (on the basis of the experimental evidence) without any attempt to provide deeper explanation. It was postulated that (only several examples)

- absolute space does not exist, whereas no one could deny that all objects exist and move in something that is called 'the space' (absolute space means *one* space that is common to all observers)

- absolute motion does not exist, whereas object move in 'the space'

- absolute time does not exist; everyone was supposed to accept the fact that that mysterious result followed from the experiment and no more questions should be asked

- the speed of light was constant no matter how the source and the observer moved, and again everyone had to just accept this paradoxical result since it was what the experimental evidence showed

- the length of bodies along the line of their motion contracts, but the reality of that relativistic contraction was another mystery since no deformation and no force were causing the shortening of the length.

Shortly after Einstein published his special relativity his mathematics professor, Hermann Minkowski, delivered three lectures (before his untimely departure from this strange world on January 12, 1909) which dramatically changed the way we understand the world [4]. He realized that all the experimental evidence supporting Galileo's principle of relativity (absolute motion with constant velocity cannot be discovered through mechanical experiments), including the failed attempts by Michelson and Morley to detect the Earth's motion by electromagnetic experiments involving light beams (light is an electromagnetic wave), carried a stunning message – all mechanical and electromagnetic experiments to discover uniform motion with respect to the absolute space (i.e. absolute motion) fail since observers in relative motion have different spaces and times, which is possible in a four-dimensional world whose fourth dimension is formed by *all* moments of time. On September 21, 1908 in his lecture "Space and Time" Minkowski announced the revolutionary view of space and time deduced from experimental physics by successfully decoding the profound message hidden in the failed experiments to discover absolute motion [4, p.111]:

> The views of space and time which I want to present to you arose from the domain of experimental physics, and therein lies their strength. Their tendency is radical. From now onwards space by itself and time by itself will recede completely to become mere shadows and only a type of union of the two will still stand independently on its own.

In his 1905 paper Einstein abandoned the idea of absolute time by postulating that the (real) time t of a stationary observer and the abstract mathematical time t', which Lorentz introduced calling it the *local time* of a moving observer, should be treated equally. As a mathematician it may have been easier for Minkowski (than for Einstein) to postulate that the times t and t' are equivalent and to explore the consequences of such a hypothesis. The mathematical way of thinking surely had helped Minkowski to realize that if two observers in relative motion have different times they necessarily must have different spaces as well (since space is perpendicular to time), which is impossible in a three-dimensional world, but is almost self-evident in a four-dimensional world with all moments of time as the fourth dimension. Here is how Minkowski in his own words at his lecture "Space and Time" explained how he had realized the profound *physical meaning of the relativity principle* – that the world is four-dimensional. In the case of two inertial reference frames in relative motion along their x-axes [4, p. 114] one can call t' time, but then must necessarily, in connection with this, define space by the manifold of three parameters x', y, z in which the laws of physics would then have exactly the same expressions by means of x', y, z, t' as by means of x, y, z, t. Hereafter we would then have in the world no more *the* space, but an infinite number of spaces analogously as there is an infinite number of planes in three-dimensional space. Three-dimensional geometry becomes a chapter in four-dimensional physics. You see why I said at the beginning that space and time will recede completely to become mere shadows and only a world in itself will exist.

Minkowski found that the relativity principle, which Einstein used as the first postulate of his theory of special relativity (the second postulate being the constancy of the speed of light), did not adequately represent what the experimental evidence tells us about the world and noted that "I think the word *relativity postulate* used for the requirement of invariance under the group G_c is very feeble. Since the meaning of the postulate is that through the phenomena only the four-dimensional world in space and time is given, but the projection in space and in time can still be made with certain freedom, I want to give this affirmation rather the name *the postulate of the absolute world*" [4, p. 117].

Minkowski suddenly found the answers to many questions in the four-dimensional physics of the absolute world (which we now call space-time or Minkowski spacetime), e.g. the answer to the question of why the relativity principle requires that physical phenomena be the same in all inertial reference frames. This is so because every inertial observer describes the phenomena in *exactly the same way* – in his own reference frame (i.e. in terms of his own space and time) in which he is *at rest*. Also, the answer to the question of the failure of Michelson and Morley's experiments to detect the motion of the Earth appears obvious – the Earth is at rest with respect to its space and therefore not only Michelson and Morley's but any other experiments would confirm this state of rest. As every observer always measures the velocity of light in his own (rest) space and by using his own time, the velocity of light is the same for all observers. As the absolute world is four-dimensional it becomes clear why there does not exist *one* (absolute) space and therefore why there does not exist absolute motion.

Also, time is not absolute since observers in relative motion can choose their times along different directions in spacetime. This becomes entirely clear when one takes into account that the whole history in time of an ordinary particle is entirely given in spacetime since *spacetime*

72

contains all moments of time at once. Minkowski called the 'line of life' of a particle a worldline. Fig. 5.1 shows the worldlines of four particles – three (a, b, c) moving by inertia (with constant velocity) and the fourth (d) accelerating. Three observers at rest with respect to particles a, b, and c can choose their time axes along the worldlines of these particles.

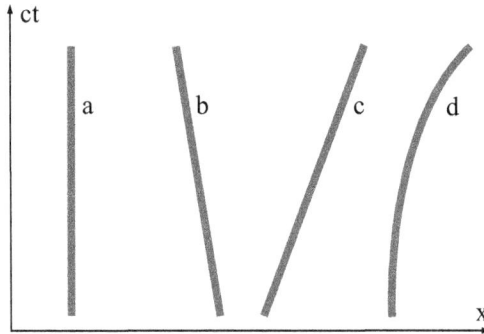

Figure 5.1: Worldlines of three particles (a, b, c) moving with constant relative velocities and of an accelerating particle (d)

Minkowski wrote [4, p. 112]: "The whole world presents itself as resolved into such worldlines, and I want to say in advance, that in my understanding the laws of physics can find their most complete expression as interrelations between these worldlines." Minkowski's absolute world essentially suggests that four-dimensional physics is in fact spacetime geometry. As important results can be deduced if this is really so, it is important to answer the fundamental question "Is the absolute four-dimensional world (spacetime) real or it is just a mathematical abstraction?" I will give again Eddington's direct answer to this question quoted in the Introduction [3, p. 803]:

> It was shown by Minkowski that all these fictitious spaces and times can be united in a single continuum of four dimensions. The question is often raised whether this four-dimensional space-time is real, or merely a mathematical construction; perhaps it is sufficient to reply that it can at any rate not be less real than the fictitious space and time which it supplants.

It is an unfortunate fact that for over a hundred years after Minkowski's lecture "Space and Time" the issue of the reality of the absolute four-dimensional world has been a controversial one, not because there are scientific reasons that question its reality, but because of the challenging implications of this forever given 'frozen' world for a number of delicate issues (one of which is the existence of free will). But should the view of the world unambiguously coming from the experimental evidence be ignored because it contradicts some of our accepted and deceivingly comfortable, but not scientifically supported views? What is even worse is that the four-dimensional view of the world not merely follows from the experimental evidence. The relativistic experimental evidence would be impossible if the world were not four-dimensional; that evidence is impossible in a three-dimensional world.

As a productive research strategy should be able to deal with such an issue let us see why this is so. The best way to see why special relativity is impossible in a three-dimensional world, assume that that is precisely the case – that the world is indeed three-dimensional and spacetime is merely an abstract concept that does not represent a real four-dimensional world.

If the world were three-dimensional, space would be absolute since there would exist only one space that would be shared by all observers in relative motion. As space constitutes a single class of simultaneous events, absolute space implies absolute simultaneity and therefore absolute time as well. All this in clear contradiction with special relativity.

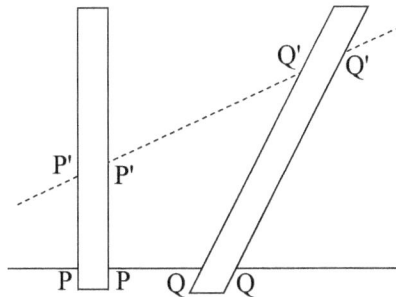

Figure 5.2: Minkowski's Fig. 1

A more detailed argument is Minkowski's own explanation of the deep physical meaning of length contraction of two bodies in relative motion. The essence of his explanation is that length contraction is a manifestation of the reality of the bodies' worldlines or rather world-tubes[1] (Minkowski called them strips). This can be best understood

---

[1]Minkowski called the entire history in time of a point a worldline, but for a

from Fig. 1 of Minkowski's paper "Space and Time" (the right-hand part of which is reproduced in Fig. 5.2 here) – length contraction would be *impossible* if the worldtubes of the two bodies, represented by the vertical and the inclined strips in Fig. 5.2, did not exist and were nothing more than abstract geometric constructions. To see this even more clearly consider only the body represented by the vertical worldtube. The three-dimensional cross-section PP, resulting from the intersection of the body's worldtube and the space of an observer at rest with respect to the body, is the body's proper length. The three-dimensional cross-section $P'P'$, resulting from the intersection of the body's worldtube and the space of an observer moving with respect to the body, is the relativistically contracted length of the body measured by that observer[2]. Minkowski stressed that "This is the meaning of the Lorentzian hypothesis of the contraction of electrons in motion" [4, p.116] and "that the Lorentzian hypothesis is completely equivalent to the new concept of space and time, which makes it much easier to understand" [4, p. 116].

By demonstrating that the length contraction of a body is a manifestation of the reality the body's worldtube (and therefore of the reality of the absolute four-dimensional world) Minkowski also showed that this effect involves no deformation and no force causing the shortening of the body's length; as seen in Fig. 5.2 the contracted body measured by the observer moving relative to the body is simply *a different cross-section $P'P'$* of the body's worldtube, which is shorter that the cross-section PP measured by the observer at rest with respect to the body. So, the length contraction effect is a nice illustration of the essence of Minkowski's four-dimensional physics – that *four-dimensional physics is spacetime geometry.*

It should be stressed that *the worldtube of the body must be real in order that length contraction be possible* because, while measuring the *same* body, the two observers in relative motion measure *two* three-dimensional bodies represented by the cross-sections PP and $P'P'$ in Fig. 5.2. This is not so surprising when one takes into account relativity of simultaneity and the fact that a spatially extended three-dimensional object is defined in terms of *simultaneity* – all parts of a body taken *simultaneously* at a given moment. If the worldtube of the body were an abstract geometric construction and what existed were a single three-dimensional body (a single class of simultaneous events) represented by the proper cross-section PP, both observers would measure the *same*

---

spatially extended body it is appropriate to use the term worldtube.

[2]The cross-section $P'P'$ only appears longer than PP because a fact of the pseudo-Euclidean geometry of spacetime is represented on the Euclidean surface of the page.

three-dimensional body of the *same* length, i.e. the *same* class of simultaneous events, which means that simultaneity would be absolute.

Although Minkowski's explanation of length contraction of a body, taken even alone, is sufficient to prove that this effect would not be possible if the body's worldtube were not real, that is, if the world were three-dimensional,[3] the same holds for all kinematical relativistic effects – those effects, and most importantly the *experiments* which confirmed them, are impossible in a three-dimensional world [6].

The *experimental proof* of the reality of the absolute four-dimensional world and of the worldtubes of macroscopic bodies should be stressed as strongly as possible since it is sometimes tempting to think that we should not take the implications of special relativity too seriously for two reasons:

- special relativity may soon be replaced by another theory that may describe the world in a different way; in fact, general relativity has already generalized special relativity and contains it as a special case

- quantum physics has already proven that elementary particles are not worldlines in spacetime, which is not surprising since the equations of motion of special relativity manifestly fail to describe the behaviour of quantum objects.

Such a temptation should be resisted since the above reasons reveal an inadequate view on the nature of physical theories that can prevent anyone who holds it from deep understanding of the world and, as a result, from making discoveries. First, as it is the *experiments*, which confirm the relativistic effects, that would be impossible in a three-dimensional world, no experimental evidence of future theories can change this fact – *experiments do not contradict one another*. Moreover, general relativity incorporates special relativity and provides further experimental evidence for the reality of spacetime. Second, by the same argument (experiments do not contradict one another), no quantum mechanical experiment can contradict the relativistic experiments that are impossible in a three-dimensional world, which therefore proved the reality of the absolute four-dimensional world and the reality of the worldtubes of *macroscopic* bodies. That elementary particles are not worldlines in spacetime only indicates *what they are not*; they might be more complex structures in spacetime such that the probabilistic behaviour of the quantum object is merely a manifestation of

---

[3]Length contraction was experimentally tested, along with time dilation, by the muon experiment in the muon reference frame (see for instance [5]).

a probabilistic distribution of the quantum object itself in the forever given spacetime (for an example see [6, Chap. 10] and the references therein) – had Minkowski lived longer he might have described such a spacetime structure by the mystical expression "predetermined probabilistic phenomena."

One of the most important features of a productive research strategy is to identify the reliable elements of our present knowledge (unambiguously extracted from the experimental and theoretical evidence) which should form a solid base for future research. Such an element of reliable knowledge is the reality of the absolute four-dimensional world and therefore the reality of the worldtubes of macroscopic bodies. Let us now briefly explore the internal logic of Minkowski's absolute world and see what some of the implications of such a view are.

We have already seen how Minkowski's realization that the relativity principle implies many times and spaces (which in turn implies that the world is four-dimensional) naturally explained why there is no absolute motion – there are many spaces, not just one absolute space with respect to which bodies move. Minkowski also generalized Newton's first law (of inertia) by pointing out that a free particle, which moves by inertia, is a straight timelike worldline in the absolute world (spacetime) and pointed out that an accelerating particle is represented by a curved worldline. Here is how he described the three states of motion of a particle (corresponding to the worldlines a, c, and d in Fig. 5.1) [4, p. 115]:

> a straight worldline parallel to the t-axis corresponds to a stationary substantial point, a straight line inclined to the t-axis corresponds to a uniformly moving substantial point, a somewhat curved worldline corresponds to a non-uniformly moving substantial point.

Minkowski found it necessary to stress that "Especially the concept of *acceleration* acquires a sharply prominent character" [4, p. 117]. This sharply prominent character of the acceleration comes from the *absolute* geometric property of the worldline of an accelerated body – the worldline of such a body is curved.

Had Minkowski's generalization of Newton's first law been carefully analyzed, several immediate consequences would have been clearly realized:

- Two pieces of reliable knowledge about an accelerating body now appear linked – an accelerating body (i) resists its acceleration, and (ii) is represented by a curved (and therefore *deformed*) world-tube. As we consider the body's worldtube to be real (based on the relativistic experimental evidence) linking these features of the accelerating body seems logically quite natural, but totally unexpected – the resistance an accelerating body offers to its acceleration can be viewed as originating from a four-dimensional stress, which arises in the deformed worldtube of the body. Had Minkowski lived longer he almost certainly would have arrived at this conclusion and would have been thrilled that his program "the laws of physics can find their most complete expression as inter-relations between these worldlines" [4, p. 112] had produced an important result – inertia originates in the body itself, or rather in the deformed worldtube of the accelerating body.

- The experimental fact that accelerated motion is detectable (due to the resistance an accelerating body offers) and therefore absolute (frame-independent) is also naturally linked to the absolute geometric property of the curved (deformed) worldtube of the accelerating body. This results finally settles the dispute about the absoluteness of acceleration dating back to Newton – *the absoluteness of acceleration merely reflects the absolute fact that the worldtube of an accelerating body is curved (deformed) and does not imply an absolute space with respect to which the body accelerates.*

- Identifying inertia (the resistance a body offers to its acceleration) with the four-dimensional stress arising in the deformed world-tube of the body also explains the difference between inertial and accelerated motion – the worldtube of a body moving by inertia is straight (not deformed) and for this reason the body offers no resistance to its inertial motion, whereas an accelerating body resists its motion since its worldtube is curved deformed. That is why inertial motion cannot be detected experimentally (what the principle of relativity states), whereas accelerated motion is experimentally observed due to the inertia (resistance) of the accelerating body.

The identification of inertia with the four-dimensional stress in the deformed worldtube of an accelerating body is also justified by the fact that the calculation of the static restoring force ($\mathbf{F}_{in}$ in Fig. 5.3) confirms that it has the form of the inertial force $\mathbf{F}_{in} = -m\mathbf{a}$ [6, Chap. 9]. As seen in Fig. 5.3 the acting force $\mathbf{F}$ which accelerates a body (i.e. deforms its worldtube depicted in the figure) is in a static equilibrium with the inertial force $\mathbf{F}_{in}$, which is the static restoring force arising the deformed worldtube of the body (exactly like the static restoring force in a deformed rod). In the ordinary three-dimensional language that equilibrium is dynamic. So the four-dimensional situation represented in Fig. 5.3 reveals the deep physical meaning of d'Alembert's principle discussed at the end of Chapter 2.

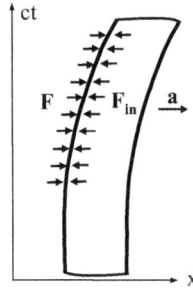

Figure 5.3: The inertia of an accelerating body (i.e. the resistance it offers to its acceleration) is caused by the four-dimensional stress arising in the deformed worldtube of the body.

Since inertial mass is defined as the measure of the resistance a body offers to its acceleration it is now clear that a body's inertial mass is linked to the four-dimensional stress in the body's deformed worldtube.[4] I think this situation provides additional insight into the nature of relativistic mass especially in view of some recent attempts to deny the relativistic increase of mass (see Appendix B).[5] Paying particular attention to the main feature of inertia (*resistance*) captured in the accepted since Newton definition of mass demonstrates that relativistic mass is indeed one of the important results of special relativity – *as inertial mass is the measure of the resistance a body offers to its acceleration and as its acceleration is different in different inertial reference frames, the body's inertial mass cannot be the same in all frames.* In the inertial reference frame in which a body is at rest its rest (or proper) mass is an invariant, but its mass is greater in all inertial reference frames in relative motion with respect to the body. This relativistic generalization of the concept

---

[4] As an elementary particle is not a worldline in spacetime its inertia appears to originate from a similar mechanism involving deformation at the quantum scale – from the *distorted* fields which mediate the particle's interactions [6, Chap. 9]; the particle's fields are deformed by its acceleration.

[5] Appendix B "On Relativistic Mass" was published as an Appendix by the Editor in [7].

of mass is analogous to the relativistic generalization of the concept of time – proper time (the time between two readings of a clock in its rest frame) is an invariant but coordinate time (the time between the same readings of the clock measured in the inertial frames moving relative to the clock) is greater; for a more detailed discussion of relativistic mass see [6, pp. 114-116].

Regarding four-dimensional physics as spacetime geometry allows us to address the problems with Mach's views which were only stated at the end of Chapter 4.

- As we saw acceleration is absolute in spacetime (it is a curvature of a worldline in spacetime), rotation is absolute and therefore the Ptolemaic and the Copernican system are not equivalent. The planets' worldtubes are helixes around the worldtube of the Sun.

- It is also clear that rotation and translation are clearly distinct in spacetime – the worldline of a particle moving translationally is either a straight line (when the particle moves uniformly) or a curved line (when the particle accelerates translationally), whereas the worldline of a rotating particle is a helix.

- According to Mach's view if there were no other bodies in the Universe, one could not talk about the state of the motion of a single particle. In spacetime the situation is crystal clear – a single particle in the Universe is either a straight worldline (which means that the particle moves uniformly) or a curved worldline (which means that the particle accelerates).

- According to Mach the origin of inertia is non-local since all the masses in the universe are responsible for the inertial forces, whereas we saw that inertia is a *local* phenomenon in spacetime since it originates from the deformation of an accelerating body's world-tube.

References

[1] A. Einstein, *The Collected Papers of Albert Einstein*, Vol. 2, transl. by A. Beck (Princeton University Press, Princeton 1989) p 140

[2] A. Pais, *Subtle Is the Lord: The Science and the Life of Albert Einstein* (Oxford University Press, Oxford 2005) p. 139

[3] A.S. Eddington, The Relativity of Time, *Nature* **106**, pp 802-804 (17 February 1921); reprinted in: A. S. Eddington, *The Theory of*

Relativity and its Influence on Scientific Thought: Selected Works on the Implications of Relativity (Minkowski Institute Press, Montreal 2015) pp. 27-30, p. 30

[4] H. Minkowski, Space and Time. New translation in: H. Minkowski, *Space and Time: Minkowski's Papers on Relativity* (Minkowski Institute Press, Montreal 2012) pp. 111–125 (http://minkowskiinstitute.org/mip/); included in this book as Appendix A

[5] G.F.R. Ellis and R.M. Williams, *Flat and Curved Space Times* (Oxford University Press, Oxford 1988) p. 104.

[6] V. Petkov, *Relativity and the Nature of Spacetime*, 2nd ed. (Springer, Heidelberg 2009) Chap. 5

[7] A. Einstein, *Relativity*, edited by V. Petkov (Minkowski Institute Press, Montreal 2018)

6. Inertia and gravitation in curved spacetime

As we saw at the end of Chapter 3 there were two problems with Newton's notion of gravitational force. A rigorous conceptual analysis of Newton's gravitational theory could have revealed them, long before Einstein. The first of those problems was realized by Einstein most probably in November 1907 and this insight set him on the path toward his theory of general relativity (quoted from [3]):

> I was sitting in a chair in the patent office at Bern when all of a sudden a thought occurred to me: "If a person falls freely he will not feel his own weight." I was startled. This simple thought made a deep impression on me. It impelled me toward a theory of gravitation.

Einstein was so impressed by this insight that he called it the "happiest thought" of his life [3]. Then Einstein needed eight years to arrive at his general relativity. This is an extremely short period for such a revolutionary theory. However, such a thinker like Einstein could have discovered the new theory of gravitation much sooner. I think he was delayed by the unproductive doctrine of relativism he shared at the time he created special relativity and later when he struggled with the theory of gravity. That doctrine is responsible for the inadequate names of his two theories of *relativity* – there is nothing fundamentally relative either in special relativity or in general relativity (all relative quantities in both theories are possible *only* due the existence of the *absolute* spacetime). Perhaps relativism had been behind Einstein's initial hostile reaction at Minkowski's four-dimensional physics. Sommerfeld's recollection of what Einstein said on one occasion provides an indication of his initial attitude toward Minkowski's work: "Since the mathematicians have invaded the relativity theory, I do not understand it myself any more" [4].

However, later Einstein adopted Minkowski's view of spacetime and his four-dimensional physics and quickly completed (in 1915) perhaps the deepest physical theory of all time – his general relativity which regards gravity not as a force but as a manifestation of the non-Euclidean geometry of spacetime (the curvature of spacetime is caused by the presence of massive bodies).

Here I would like to demonstrate what a research strategy, based on the reliable knowledge that the world is four-dimensional and that four-dimensional physics is in fact spacetime geometry, might have produced, if Einstein used it together with his happiest thought. Let us imagine that Einstein examined his happiest thought – a falling body that does not feel the force of gravity ("If a person falls freely he will not feel his own weight") – and realized that *a falling body does not resist its fall* since a body resists ("feels") its motion only if it is subject to a force (recall that by Newton's second law a force is only needed to overcome the resistance the body offers to its acceleration). Imagine also that Einstein carefully analyzed Minkowski's view of the Universe as an absolute four-dimensional world and arrived at the results discussed in Chapter 5.

Then the path to the idea that gravitational phenomena are manifestations of the curvature of spacetime would have been open to him – the experimental fact that a falling body accelerates (which means that its worldtube is curved), but offers no resistance to its acceleration (which means that its worldtube is not deformed) can be explained only if the worldtube of a falling body is *both curved and not deformed*, which is impossible in the flat Minkowski spacetime where a curved worldtube is always deformed. Such a worldtube can exist only in a non-Euclidean spacetime whose "straight" worldlines (called geodesics) are naturally curved due to the spacetime curvature, but are not deformed. This insight provides an unexpected answer to the question in the last sentence of Chapter 3 (How can a falling body move by inertia, if it accelerates?) – as there are no straight and parallel worldlines in curved spacetime, the geodesic worldlines of two bodies moving by inertia will either converge or diverge (called geodesic deviation) and as a result it will *appear* that the bodies accelerate toward or away of each other).

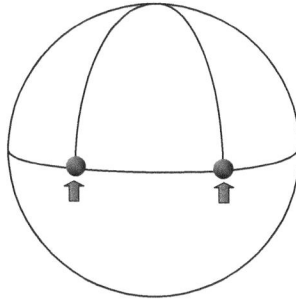

Figure 6.1: Gravitational "attraction" in two dimensional spacetime

To see better how a worldline can be curved but not deformed consider the meridians and the equator on a globe as shown in Fig. 6.1. As there are no straight lines on the curved surface of a sphere, the analog of straight lines are the geodesics (the great circles obtained by the intersection of the sphere and a plane passing through the sphere's center), i.e. the meridians and the equator in the case of the globe. They are naturally curved due to the curvature of the globe's surface, but are not deformed. If we try to curve a meridian further, we deform it.

Einstein would have realized that Minkowski's program "the laws of physics can find their most complete expression as interrelations between these worldlines" [5, p. 112] had produced another stunning result – gravitational phenomena are not caused by a force, but are merely effects due to the curvature of spacetime.

To see that gravitational phenomena are indeed geometrical effects on a curved spacetime, consider two balls on the equator as shown in Fig. 6.1, which start to move upward in directions that are perpendicular to the equator. As the balls recede from the equator they start to approach each other. If we are unaware that the balls move on the surface of a sphere and implicitly assume that they move on a plane, we will explain the shortening of the distance between them as caused by a force of attraction between the balls. This is how Newton explained the gravitational phenomena (attraction between bodies). However, when we take into account the real situation – that the balls are on the curved surface of a sphere – it is self-evident that no force is acting on them since the balls move by inertia on the curved surface; the reason they approach each other is that there are no parallel lines on the surface of the sphere.

In fact, this explanation is a bit simplified since it considered the surface of an ordinary globe and regarded the two meridians, along which the balls moved, as the *trajectories*[1] of the balls. The real situation depicted in Fig. 6.1 represents a two dimensional curved spacetime and the meridians are the *worldlines* of the balls. The two worldlines converge which in the ordinary three-dimensional language means that the balls approach each other.

Let us now return to the falling body. Such a body is represented by a geodesic (i.e. not deformed) worldtube, which means that the body moves by inertia, i.e. without resisting its fall. This fact is captured in the geodesic hypothesis in general relativity, which states that the worldline of a free particle is a timelike *geodesic* in spacetime. This hypothesis[2] is "a natural generalization of Newton's first law" [6], that is, "a mere extension of Galileo's law of inertia to curved spacetime" [7]. This means that *in general relativity a particle, whose worldline is geodesic, is a free particle which moves by inertia.*

When a falling body reaches the ground it is prevented from falling (i.e. from moving by inertia) and *resists* that change by exerting an *inertial* force on the ground, which has been traditionally called gravitational force. That the force with which the falling body resists the change in its inertial motion is inertial, is nicely demonstrated when we consider the worldtube of the body when it hits the ground. The worldtube of the body (when it is on the Earth's surface) is *deformed*, exactly like the worldtube of an accelerating body is deformed, and therefore the restoring static force that is caused by the deformation is inertial like the restoring static force originating from the deformed worldtube of the accelerating body. The mass of the body, when at rest on the ground, has been traditionally called passive gravitational mass, but we see that it is inertial mass – it is the measure of the resistance the body offers when prevented from moving by inertia while falling.

As we see it turns out that the inexplicable equivalence of the inertial mass and the passive gravitational mass (and the equivalence of the inertial and gravitational forces), which Einstein merely postulated as the equivalence principle, found a natural explanation: inertial and gravitational masses and forces are equivalent since they are the same thing – they are all inertial. Minkowski's program regarding physics as spacetime geometry again helped us to find another deep explanation –

---

[1]A trajectory of a moving body is the projection of its worldline on space.

[2]The fact that a falling body does not resist its fall, which means that it moves by inertia, is called hypothesis by tradition; this fact is experimentally confirmed – a falling accelerometer, for example, reads zero resistance since it measures acceleration through resistance.

that of the equivalence principle.

The example with the falling body also nicely explains the status of acceleration in general relativity (i.e. in curved spacetime). A falling body accelerates (while moving by inertia) but that acceleration is *apparent* since it is caused by the curvature of spacetime induced by the Earth's mass – the geodesic worldline of the Earth's center and the body's geodesic worldtube converge toward each other (there are no straight and parallel worldlines in curved spacetime) which we perceive as a body falling toward the Earth's surface.[3] The apparent acceleration, caused by the geodesic deviation of the body's geodesic worldtube and the geodesic worldline of the Earth's center, is relative and involves no deformation of the falling body's worldtube. By contrast, the worldtube of the body, when it is at rest on the Earth's surface, is *deformed* which indicates that the body's acceleration is absolute. Indeed, the body resists its being prevented from moving by inertia while falling by exerting a real inertial force on the ground.

The status of acceleration in flat and curved spacetime is the same – in both cases the acceleration of a body is *absolute* if its worldtube is *deformed*. In curved spacetime there is a second acceleration – apparent (relative) acceleration (caused by geodesic deviation) – which is a manifestation of the spacetime curvature. All gravitational accelerations (of falling bodies and of planets orbiting the Sun) are such apparent accelerations through which the non-Euclidean geometry of spacetime manifests itself.

The example with the falling body makes it possible to clarify an-

---

[3] As the falling body actually moves by inertia its apparent acceleration is exactly like the apparent acceleration of a ball moving by inertia but described in an accelerating elevator as depicted in Fig. 2.5 and explained there (Chapter 2). To see clearly why this is so, consider an identical elevator placed on the Earth's surface. The worldtube of the elevator is deformed by the surface of the Earth, which is preventing it from moving by inertia toward the center of the Earth (i.e., preventing it from following its geodesic path; therefore, the elevator, which is stationary on the Earth's surface, is subject to an absolute acceleration; keep in mind that in spacetime physics a body is subject to an absolute acceleration when its worldtube is deformed (i.e., not geodesic); rather, it is spacetime physics which *revealed* that an absolute acceleration of a body means that the body's worldtube is deformed. A ball released in the elevator will move by inertia (non-resistantly) while falling since its worldtube is geodesic. So, in both cases – a ball falling in an accelerating elevator (in flat spacetime) and in an elevator at rest on the Earth's surface (in curved spacetime) – moves by inertia (its worldtube is geodesic), whereas the elevator is absolutely accelerating (its worldtube is deformed, i.e., not geodesic). The equivalence of physical phenomena occurring in an accelerating elevator and in an elevator at rest on the Earth's surface, closely examined by Einstein, was postulated by him as the equivalence principle and led him to the identification of gravitational phenomena with the non-Euclidean geometry of spacetime.

other misunderstanding ultimately coming from Mach – that the shape a geodesic worldtube is determined by all masses in the Universe and therefore when a body is deviated from its geodesic path the inertial force with which the body resists the change in its inertial motion is determined by all the masses and is therefore gravitational in origin. That all masses in the Universe determine the global curvature of the Universe is undeniable, but the claim that the inertial force arising in the deformed worldtube of a body is gravitational is simply untrue. The shape of the geodesic worldtube of a body falling toward the Earth is determined *predominantly* by the spacetime curvature induced by the Earth's mass. There are small contributions from the Moon and the Sun and practically zero contributions from the distant masses in the Universe. When the body is at rest on the Earth's surface its worldtube is deformed and the resulting inertial force (traditionally called gravitational force or the body's weight) originates from the four-dimensional stress in the bodies own worldtube; that is, the body's inertia originates in the body itself. As the shape of the falling body's geodesic worldtube is almost entirely determined by the spacetime curvature caused by the Earth's mass, the question of whether the body's inertial force (exerted on the ground) originates completely from the body or the Earth's mass has a contribution to that force will remain open until we understand how the Earth curves the spacetime around its worldtube (in order to determine whether spacetime transmits some interaction between the Earth and the body; see Appendix D).

Had Einstein followed Minkowski's program that four-dimensional physics is nothing more than spacetime geometry, he would have certainly arrived at the conclusion that gravitational phenomena are not caused by gravitational interaction since they are fully explained in the theory without the need to assume the existence of gravitational interaction: what has the appearance of gravitational attraction involves only *inertial* (*interaction-free*) motion and is merely a manifestation of the curvature of spacetime. For example, the planets are free bodies which move by inertia (since their worldlines are geodesics) and as such they do not interact in any way with the Sun because *inertial motion does not imply any interaction*. However, a complete understanding of gravitational phenomena requires one more step – to resolve the major open question of how matter curves spacetime, and not to try to quantize gravity since there is no gravitational interaction to quantize.

However Einstein followed a different path and even such a great thinker was unable to free himself completely from the deceivingly overwhelming evidence that gravity is interaction and that gravitational energy and momentum are involved in the gravitational phenomena (for a discussion on the existence of gravitational energy and on quantum gravity see Appendix D).

Einstein's own (not very clear) view of whether gravity is a manifestation of the non-Euclidean geometry of spacetime, might have contributed to the persisting confusion about the nature of gravitation and gravitational energy. As Lehmkuhl [8] has recently discussed, Einstein did not seem to believe that general relativity geometrized gravitation: "I do not agree with the idea that the general theory of relativity is geometrizing Physics or the gravitational field" [9]. This quote appears to suggest that Einstein looked at the mathematical formalism of general relativity as pure mathematics and regarded gravitation as a physical interaction involving exchange of gravitational energy and momentum. Einstein made the gigantic step to identify gravity with the curvature of spacetime, but it seems even a genius could have doubts about his own revolutionary discovery.

References

[1] A.S. Eddington, The Relativity of Time, *Nature* **106**, 802-804 (17 February 1921); reprinted in: A. S. Eddington, *The Theory of Relativity and its Influence on Scientific Thought: Selected Works on the Implications of Relativity* (Minkowski Institute Press, Montreal 2015) pp. 27-30, p. 30

[2] A.S. Eddington, *The Mathematical Theory of Relativity* (Minkowski Institute Press, Montreal 2016) p. 233

[3] A. Pais, *Subtle Is the Lord: The Science and the Life of Albert Einstein* (Oxford, Oxford University Press 2005) p. 179

[4] A. Sommerfeld, To Albert Einstein's Seventieth Birthday. In: *Albert Einstein: Philosopher-Scientist*. P. A. Schilpp, ed., 3rd ed. (Open Court, Illinois 1969) pp. 99-105, p. 102

[5] H. Minkowski, Space and Time. New translation in: H. Minkowski, *Space and Time: Minkowski's Papers on Relativity* (Minkowski Institute Press, Montreal 2012) pp. 111–125 (http://minkowskiinstitute.org/mip/); included in this book as Appendix A

[6] J. L. Synge, *Relativity: the general theory.* (Nord-Holand, Amsterdam 1960) p. 110

88

[7] W. Rindler, *Relativity: Special, General, and Cosmological* (Oxford University Press, Oxford 2001) p. 178

[8] D. Lehmkuhl, Why Einstein did not believe that General Relativity geometrizes gravity. *Studies in History and Philosophy of Physics*, Volume 46, May 2014, pp. 316-326

[9] A letter from Einstein to Lincoln Barnett from June 19, 1948; quoted in [8]

Appendix A:
H. Minkowski, Space and Time

Lecture given at the 80th Meeting of the Natural Scientists in Cologne on
September 21, 1908 (H. Minkowski, Raum und Zeit, *Physikalische
Zeitschrift* **10** (1909) S. 104-111)
New translation by V. Petkov, reprinted here from: H. Minkowski, *Space
and Time: Minkowski's Papers on Relativity*, Edited by V. Petkov
(Minkowski Institute Press, Montreal 2012)

Gentlemen! The views of space and time which I want to present
to you arose from the domain of experimental physics, and therein lies
their strength. Their tendency is radical. From now onwards space by
itself and time by itself will recede completely to become mere shadows
and only a type of union of the two will still stand independently on its
own.

I.

I want to show first how to move from the currently adopted mechanics through purely mathematical reasoning to modified ideas about
space and time. The equations of Newtonian mechanics show a twofold
invariance. First, their form is preserved when subjecting the specified
spatial coordinate system to *any change of position*; second, when it
changes its state of motion, namely when any *uniform translation* is
impressed upon it; also, the zero point of time plays no role. When one
feels ready for the axioms of mechanics, one is accustomed to regard the
axioms of geometry as settled and probably for this reason those two
invariances are rarely mentioned in the same breath. Each of them represents a certain group of transformations for the differential equations
of mechanics. The existence of the first group can be seen as reflecting
a fundamental characteristic of space. One always tends to treat the
second group with disdain in order to unburden one's mind that one
can never determine from physical phenomena whether space, which is

assumed to be at rest, may not after all be in uniform translation. Thus these two groups lead completely separate lives side by side. Their entirely heterogeneous character may have discouraged any intention to compose them. But it is the composed complete group as a whole that gives us to think.

We will attempt to visualize the situation graphically. Let x, y, z be orthogonal coordinates for space and let t denote time. The objects of our perception are always connected to places and times. No one has noticed a place other than at a time and a time other than at a place. However I still respect the dogma that space and time each have an independent meaning. I will call a point in space at a given time, i.e. a system of values x, y, z, t a *worldpoint*. The manifold of all possible systems of values x, y, z, t will be called the *world*. With a hardy piece of chalk I can draw four world axes on the blackboard. Even *one* drawn axis consists of nothing but vibrating molecules and also makes the journey with the Earth in the Universe, which already requires sufficient abstraction; the somewhat greater abstraction associated with the number 4 does not hurt the mathematician. To never let a yawning emptiness, let us imagine that everywhere and at any time something perceivable exists. In order not to say matter or electricity I will use the word substance for that thing. We focus our attention on the substantial point existing at the worldpoint x, y, z, t and imagine that we can recognize this substantial point at any other time. A time element dt may correspond to the changes dx, dy, dz of the spatial coordinates of this substantial point. We then get an image, so to say, of the eternal course of life of the substantial point, a curve in the world, a *worldline*, whose points can be clearly related to the parameter t from $-\infty$ to $+\infty$. The whole world presents itself as resolved into such worldlines, and I want to say in advance, that in my understanding the laws of physics can find their most complete expression as interrelations between these worldlines.

Through the concepts of space and time the x, y, z-manifold $t = 0$ and its two sides $t > 0$ and $t < 0$ fall apart. If for simplicity we hold the chosen origin of space and time fixed, then the first mentioned group of mechanics means that we can subject the x, y, z-axes at $t = 0$ to an arbitrary rotation about the origin corresponding to the homogeneous linear transformations of the expression

$$x^2 + y^2 + z^2.$$

The second group, however, indicates that, also without altering the expressions of the laws of mechanics, we may replace

$$x, y, z, t \quad \text{by} \quad x - \alpha t, \; y - \beta t, \; z - \gamma t, \; t,$$

where α, β, γ are any constants. The time axis can then be given a completely arbitrary direction in the upper half of the world $t > 0$. What has now the requirement of orthogonality in space to do with this complete freedom of choice of the direction of the time axis upwards?

To establish the connection we take a positive parameter c and look at the structure

$$c^2 t^2 - x^2 - y^2 - z^2 = 1.$$

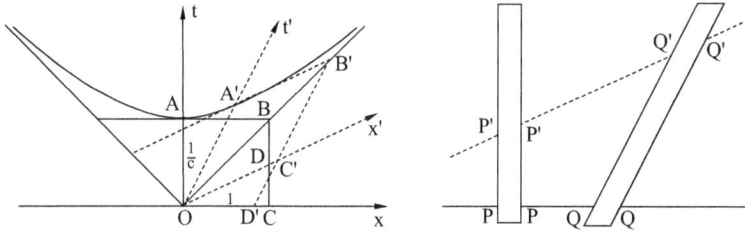

Fig. 1

It consists of two sheets separated by $t = 0$ by analogy with a two-sheeted hyperboloid. We consider the sheet in the region $t > 0$ and we will now take those homogeneous linear transformations of x, y, z, t in four new variables x', y', z', t' so that the expression of this sheet in the new variables has the same form. Obviously, the rotations of space about the origin belong to these transformations. A full understanding of the rest of those transformations can be obtained by considering such among them for which y and z remain unchanged. We draw (Fig. 1) the intersection of that sheet with the plane of the x- and the t-axis, i.e. the upper branch of the hyperbola $c^2 t^2 - x^2 = 1$ with its asymptotes. Further we draw from the origin O an arbitrary radius vector OA' of this branch of the hyperbola; then we add the tangent to the hyperbola at A' to intersects the right asymptote at B'; from $OA'B'$ we complete the parallelogram $OA'B'C'$; finally, as we will need it later, we extend $B'C'$ so that it intersects the x-axis at D'. If we now regard OC' and OA' as axes for new coordinates x', t', with the scale units $OC' = 1$, $OA' = 1/c$, then that branch of the hyperbola again obtains the expression $ct'^2 - x'^2 = 1$, $t' > 0$, and the transition from x, y, z, t to x', y', z', t' is one of the transformations in question. These transformations plus the arbitrary displacements of the origin of space and time constitute a group of transformations which still depends on the parameter c and which I will call G_c.

If we now increase c to infinity, so $1/c$ converges to zero, it is clear from the figure that the branch of the hyperbola leans more and more towards the x-axis, that the angle between the asymptotes becomes greater, and in the limit that special transformation converts to one where the t'-axis may be in any upward direction and x' approaches x ever more closely. By taking this into account it becomes clear that the group G_c in the limit $c = \infty$, that is the group G_∞, is exactly the complete group which is associated with the Newtonian mechanics. In this situation, and since G_c is mathematically more understandable than G_∞, there could have probably been a mathematician with a free imagination who could have come up with the idea that at the end natural phenomena do not actually possess an invariance with the group G_∞, but rather with a group G_c with a certain finite c, which is *extremely great* only in the ordinary units of measurement. Such an insight would have been an extraordinary triumph for pure mathematics. Now mathematics expressed only staircase wit here, but it has the satisfaction that, due to its happy antecedents with their senses sharpened by their free and penetrating imagination, it can grasp the profound consequences of such remodelling of our view of nature.

I want to make it quite clear what the value of c will be with which we will be finally dealing. c is the *velocity of the propagation of light in empty space*. To speak neither of space nor of emptiness, we can identify this magnitude with the ratio of the electromagnetic to the electrostatic unit of the quantity of electricity.

The existence of the invariance of the laws of nature with respect to the group G_c would now be stated as follows:

From the entirety of natural phenomena, through successively enhanced approximations, it is possible to deduce more precisely a reference system x, y, z, t, space and time, by means of which these phenomena can be then represented according to certain laws. But this reference system is by no means unambiguously determined by the phenomena. *One can still change the reference system according to the transformations of the above group G_c arbitrarily without changing the expression of the laws of nature in the process.*

For example, according to the figure depicted above one can call t' time, but then must necessarily, in connection with this, define space by the manifold of three parameters x', y, z in which the laws of physics would then have exactly the same expressions by means of x', y, z, t' as by means of x, y, z, t. Hereafter we would then have in the world no more *the* space, but an infinite number of spaces analogously as there is an infinite number of planes in three-dimensional space. Three-dimensional geometry becomes a chapter in four-dimensional physics. You see why I said at the beginning that space and time will recede completely to become mere shadows and only a world in itself will exist.

II.

Now the question is, what circumstances force us to the changed view of space and time, does it actually never contradict the phenomena, and finally, does it provide advantages for the description of the phenomena?

Before we discuss these questions, an important remark is necessary. Having individualized space and time in some way, a straight worldline parallel to the t-axis corresponds to a stationary substantial point, a straight line inclined to the t-axis corresponds to a uniformly moving substantial point, a somewhat curved worldline corresponds to a non-uniformly moving substantial point. If at any worldpoint x, y, z, t there is a worldline passing through it and we find it parallel to any radius vector OA' of the previously mentioned hyperboloidal sheet, we may introduce OA' as a new time axis, and with the thus given new concepts of space and time, the substance at the worldpoint in question appears to be at rest. We now want to introduce this fundamental axiom:

With appropriate setting of space and time the substance existing at any worldpoint can always be regarded as being at rest.

This axiom means that at every worldpoint[4] the expression

$$c^2 dt^2 - dx^2 - dy^2 - dz^2$$

is always positive, which is equivalent to saying that any velocity v is always smaller than c. Then c would be an upper limit for all substantial velocities and that is precisely the deeper meaning of the quantity c. In this understanding the axiom is at first glance slightly displeasing. It should be noted, however, that a modified mechanics, in which the square root of that second order differential expression enters, is now gaining ground, so that cases with superluminal velocity will play only such a role as that of figures with imaginary coordinates in geometry.

The *impulse* and true motivation for *accepting the group* G_c came from noticing that the differential equation for the propagation of light waves in the empty space possesses that group G_c[5]. On the other hand, the concept of a rigid body has meaning only in a mechanics with the

---

[4] *Editor's note:* Minkowski means at every worldpoint along the worldline of the substance.

[5] An important application of this fact can already be found in W. Voigt, Göttinger Nachrichten, 1887, S. 41.

group G_∞. If one has optics with G_c, and if, on the other hand, there were rigid bodies, it is easy to see that *one* t-direction would be distinguished by the two hyperboloidal sheets corresponding to G_c and G_∞, and would have the further consequence that one would be able, by using appropriate rigid optical instruments in the laboratory, to detect a change of phenomena at various orientations with respect to the direction of the Earth's motion. All efforts directed towards this goal, especially a famous interference experiment of Michelson had, however, a negative result. To obtain an explanation, H. A. Lorentz made a hypothesis, whose success lies precisely in the invariance of optics with respect to the group G_c. According to Lorentz every body moving at a velocity v must experience a reduction in the direction of its motion namely in the ratio

$$1 : \sqrt{1 - \frac{v^2}{c^2}}.$$

This hypothesis sounds extremely fantastical. Because the contraction is not to be thought of as a consequence of resistances in the ether, but merely as a gift from above, as an accompanying circumstance of the fact of motion.

I now want to show on our figure that the Lorentzian hypothesis is completely equivalent to the new concept of space and time, which makes it much easier to understand. If for simplicity we ignore y and z and think of a world of one spatial dimension, then two strips, one upright parallel to the t-axis and the other inclined to the t-axis (see Fig. 1), are images for the progression in time of a body at rest and a body moving uniformly, where each preserves a constant spatial dimension. OA' is parallel to the second strip, so we can introduce t' as time and x' as a space coordinate and then it appears that the second body is at rest, whereas the first – in uniform motion. We now assume that the first body has length l when considered at rest, that is, the cross section PP of the first strip and the x-axis is equal to $l \cdot OC$, where OC is the measuring unit on the x-axis, and, on the other hand, that the second body has the same length l when *regarded at rest*; then the latter means that the cross-section of the second strip *measured parallel to the x'-axis* is $Q'Q' = l \cdot OC'$. We have now in these two bodies images of two *equal* Lorentz electrons, one stationary and one uniformly moving. But if we go back to the original coordinates x, t, we should take as the dimension of the second electron the cross section QQ of its associated strip *parallel to the x-axis*. Now as $Q'Q' = l \cdot OC'$, it is obvious that $QQ = l \cdot OD'$. If dx/dt for the second strip is $= v$, an easy calculation gives $OD' = OC \cdot \sqrt{1 - \frac{v^2}{c^2}}$, therefore also $PP : QQ = 1 : \sqrt{1 - \frac{v^2}{c^2}}$. This is the meaning of the Lorentzian hypothesis of the contraction of electrons in motion. Regarding, on the other hand, the second electron as being at rest, that is, adopting the reference system x', t', the length of the first electron will be the cross section $P'P'$ of its strip parallel to OC', and we would find the first electron shortened with respect to the second in exactly the same proportion; from the figure we also see that

$$P'P' : Q'Q' = OD : OC' = OD' : OC = QQ : PP.$$

Lorentz called t', which is a combination of x and t, *local time* of the uniformly moving electron, and associated a physical construction with this concept for a better understanding of the contraction hypothesis. However, it is to the credit of A. Einstein[6] who first realized clearly that the time of one of the electrons is as good as that of the other, i.e. that t and t' should be treated equally. With this, time was deposed from its status as a concept unambiguously determined by the phenomena. The concept of space was shaken neither by Einstein nor by Lorentz, maybe because in the above-mentioned special transformation, where the plane of x', t' coincides with the plane x, t, an interpretation is possible as if the x-axis of space preserved its position. To step over the concept of space in such a way is an instance of what can be achieved only due to the audacity of mathematical culture. After this further step, which is indispensable for the true understanding of the group G_c, I think the word *relativity postulate* used for the requirement of invariance under the group G_c is very feeble. Since the meaning of the postulate is that through the phenomena only the four-dimensional world in space and time is given, but the projection in space and in time can still be made with certain freedom, I want to give this affirmation rather the name *the postulate of the absolute world* (or shortly the world postulate).

III.

Through the world postulate an identical treatment of the four identifying quantities x, y, z, t becomes possible. I want to explain now how, as a result of this, we gain more understanding of the forms under which the laws of physics present themselves. Especially the concept of *acceleration* acquires a sharply prominent character.

I will use a geometric way of expression, which presents itself immediately when one implicitly ignores z in the triple x, y, z. An arbitrary worldpoint O can be taken as the origin of space-time. The *cone*

---

[6]A. Einstein, Annalen der Physik 17 (1905), S. 891; Jahrbuch der Radioaktivität und Elektronik 4 (1907), S. 411.

$$c^2 t^2 - x^2 - y^2 - z^2 = 0$$

with O as the apex (Fig. 2) consists of two parts, one with values $t < 0$, the other one with values $t > 0$.

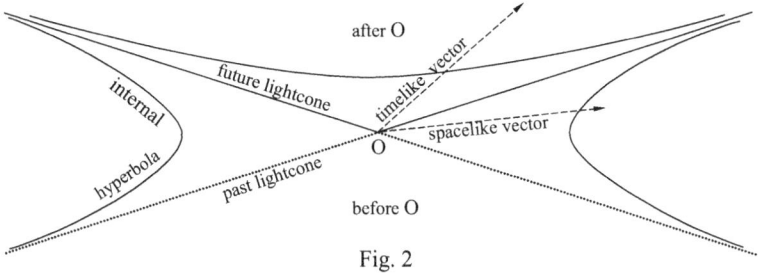

Fig. 2

The first, the *past lightcone of* O, consists, we can say, of all world-points which "send light to O", the second, the *future lightcone of* O, consists of all worldpoints which "receive light from O"[7]. The area bounded solely by the past lightcone may be called *before* O, whereas the area bounded solely by the future lightcone – *after* O. Situated after O is the already considered hyperboloidal sheet

$$F = c^2 t^2 - x^2 - y^2 - z^2 = 1, \ t > 0$$

The area *between the cones* is filled with the one-sheeted hyperboloidal structures

$$-F = x^2 + y^2 + z^2 - c^2 t^2 = k^2$$

for all constant positive values of k^2. Essential for us are the hyperbolas with O as the center, located on the latter structures. The individual branches of these hyperbolas may be briefly called *internal hyperbolas with center* O. Such a hyperbola would be thought of as the world-line of a substantive point, which represents its motion that increases asymptotically to the velocity of light c for $t = -\infty$ and $t = +\infty$.

---

[7] *Editor's and translator's note:* I decided to translate the words *Vorkegel* and *Nachkegel* as *past lightcone* and *future lightcone*, respectively, for two reasons. First, this translation reflects the essence of Minkowski's idea – (i) all worldpoints on the past lightcone "send *light* to O", which means that they all can influence O and therefore lie in the past of O; (ii) all worldpoints on the future lightcone "receive *light* from O", which means that they all can be influenced by O and therefore lie in the *future* of O. Second, the terms *past lightcone* and *future lightcone* are now widely accepted in spacetime physics.

If we now call, by analogy with vectors in space, a directed line in the manifold x, y, z, t a vector, we have to distinguish between the *timelike* vectors with directions from O to the sheet $+F = 1, t > 0$, and the *spacelike* vectors with directions from O to $-F = 1$. The time axis can be parallel to any vector of the first kind. Every worldpoint between the future lightcone and the past lightcone of O can be regarded, by a choice of the reference system, as *simultaneous* with O as well as *earlier* than O or *later* than O. Each worldpoint within the past lightcone of O is necessarily always earlier than O, each worldpoint within the future lightcone is necessarily always later than O. The transition to the limit $c = \infty$ would correspond to a complete folding of the wedge-shaped section between the cones into the flat manifold $t = 0$. In the figures this section is intentionally made with different widths.

We decompose any vector, such as that from O to x, y, z, t into four *components* x, y, z, t. If the directions of two vectors are, respectively, that of a radius vector OR from O to one of the surfaces $\mp F = 1$, and that of a tangent RS at the point R on the same surface, the vectors are called *normal* to each other. Accordingly,

$$c^2 t t_1 - x x_1 - y y_1 - z z_1 = 0$$

is the condition for the vectors with components x, y, z, t and x_1, y_1, z_1, t_1 to be normal to each other.

The *measuring units* for the *magnitudes* of vectors in different directions may be fixed by assigning to a spacelike vector from O to $-F = 1$ always the magnitude 1, and to a timelike vector from O to $+F = 1, t > 0$ always the magnitude $1/c$.

Let us now imagine a worldpoint $P(x, y, z, t)$ through which the worldline of a substantial point is passing, then the magnitude of the timelike vector dx, dy, dz, dt along the line will be

$$d\tau = \frac{1}{c} \sqrt{c^2 dt^2 - dx^2 - dy^2 - dz^2}.$$

The integral $\int d\tau = \tau$ of this magnitude, taken along the worldline from any fixed starting point P_0 to the variable end point P, we call the *proper time* of the substantial point at P. On the worldline we consider x, y, z, t, i.e. the components of the vector OP, as functions of the proper time τ; denote their first derivatives with respect to τ by $\dot{x}, \dot{y}, \dot{z}, \dot{t}$; their second derivatives with respect to τ by $\ddot{x}, \ddot{y}, \ddot{z}, \ddot{t}$, and call the corresponding vectors, the derivative of the vector OP with respect to τ the *velocity vector at* P and the derivative of the velocity vector with respect to τ the *acceleration vector at* P. As

98

$$c^2\dot{t}^2 - \dot{x}^2 - \dot{y}^2 - \dot{z}^2 = c^2$$

it follows that

$$c^2\dot{t}\ddot{t} - \dot{x}\ddot{x} - \dot{y}\ddot{y} - \dot{z}\ddot{z} = 0,$$

i.e. the velocity vector is the timelike vector of magnitude 1 in the direction of the worldline at P, and the acceleration vector at P is normal to the velocity vector at P, so it is certainly a spacelike vector.

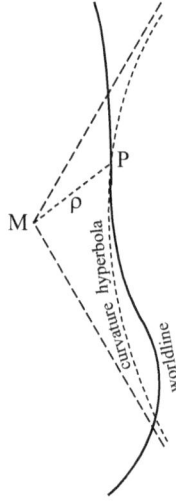

Fig. 3

Now there is, as is easily seen, a specific branch of the hyperbola, which has three infinitely adjacent points in common with the worldline at P, and whose asymptotes are generators of a past lightcone and a future lightcone (see Fig. 3). This branch of the hyperbola will be called the *curvature hyperbola* at P. If M is the center of this hyperbola, we have here an internal hyperbola with center M. Let ρ be the magnitude of the vector MP, *so we recognize the acceleration vector at P as the vector in the direction MP of magnitude c^2/ρ.*

If $\ddot{x}, \ddot{y}, \ddot{z}, \ddot{t}$ are all zero, the curvature hyperbola reduces to the straight line touching the worldline at P, and we should set $\rho = \infty$.

IV.

To demonstrate that the adoption of the group G_c for the laws of physics never leads to a contradiction, it is inevitable to undertake a revision of all physics based on the assumption of this group. This revision has been done successfully to some extent for questions of thermodynamics and heat radiation[8], for the electromagnetic processes, and finally, with the retention of the concept of mass, for mechanics.[9]

For the latter domain, the question that should be raised above all is: When a force with the spatial components X, Y, Z acts at a world-point $P(x, y, z, t)$, where the velocity vector is $\dot{x}, \dot{y}, \dot{z}, \dot{t}$, as what force this force should be interpreted for any change of the reference system? Now there exist some proven approaches to the ponderomotive force in the electromagnetic field in cases where the group G_c is undoubtedly permissible. These approaches lead to the simple rule: *When the reference system is changed, the given force transforms into a force in the new space coordinates in such a way that the corresponding vector with the components*

$$\dot{t}X, \quad \dot{t}Y, \quad \dot{t}Z, \quad \dot{t}T$$

remains unchanged, and where

$$T = \frac{1}{c^2}\left(\frac{\dot{x}}{\dot{t}}X + \frac{\dot{y}}{\dot{t}}Y + \frac{\dot{z}}{\dot{t}}Z\right)$$

is the work done by the force at the worldpoint divided by c^2. This vector is always normal to the velocity vector at P. Such a force vector, representing a force at P, will be called a *motive force vector* at P.

Now let the worldline passing through P represent a substantial point with constant *mechanical mass* m. The multiplied by m velocity vector at P will be called the *momentum vector at* P, and the multiplied by m acceleration vector at P will be called the *force vector of the motion at* P. According to these definitions, the law of motion for a point mass with a given force vector is:[10]

The force vector of the motion is equal to the motive force vector.

This assertion summarizes four equations for the components for the four axes, wherein the fourth can be regarded as a consequence of the first three because both vectors are from the start normal to the velocity vector. According to the above meaning of T, the fourth equation is

---

[8]M. Planck, "Zur Dynamik bewegter Systeme," Sitzungsberichte der k. preußischen Akademie der Wissenschaften zu Berlin, 1907, S. 542 (auch Annalen der Physik, Bd. 26, 1908, S. 1).

[9]H. Minkowski, "Die Grundgleichungen für die elektromagnetischen Vorgänge in bewegten Körpern", Nachrichten der k. Gesellschaft der Wissenschaft zu Göttingen, mathematisch-physikalische Klasse, 1908, S. 53 und Mathematische Annalen, Bd. 68, 1910, S. 527

[10]H. Minkowski, loc. cit., p. 107. Cf. also M. Planck, Verhandlungen der Physikalischen Gesellschaft, Bd. 4, 1906, S. 136.

undoubtedly the law of conservation of energy. The *kinetic energy* of the point mass is defined as the *component of the momentum vector along the t-axis multiplied by c^2*. The expression for this is

$$mc^2 \frac{dt}{d\tau} = \frac{mc^2}{\sqrt{1 - \frac{v^2}{c^2}}},$$

which is, the expression $\frac{1}{2}mv^2$ of Newtonian mechanics after the subtraction of the additive constant term mc^2 and neglecting magnitudes of the order $1/c^2$. The *dependence of the energy on the reference system* is manifested very clearly here. But since the t-axis can be placed in the direction of each timelike vector, then, on the other hand, the law of conservation of energy, formed for every possible reference system, already contains the whole system of the equations of motion. In the discussed limiting case $c = \infty$, this fact will retain its importance for the axiomatic structure of Newtonian mechanics and in this sense has been already noticed by J. R. Schütz[11]

From the beginning we can determine the ratio of the units of length and time in such a way that the natural limit of velocity becomes $c = 1$. If we introduce $\sqrt{-1}t = s$ instead of t, then the quadratic differential expression

$$d\tau^2 = -dx^2 - dy^2 - dz^2 - ds^2$$

becomes completely symmetric in x, y, z, s and this symmetry is carried over to any law that does not contradict the world postulate. Thus the essence of this postulate can be expressed mathematically very concisely in the mystical formula:

$$3 \cdot 10^5 \; km = \sqrt{-1} \; seconds.$$

V.

The advantages resulting from the world postulate may most strikingly be proved by indicating the effects from *an arbitrarily moving point charge* according to the Maxwell-Lorentz theory. Let us imagine the worldline of such a pointlike electron with charge e, and take on it the proper time τ from any initial point. To determine the field induced

[11]J. R. Schütz, "Das Prinzip der absoluten Erhaltung der Energie", Nachrichten der k. Gesellschaft der Wissenschaften zu Göttingen, mathematisch-physikalische Klasse, 1897, S. 110.

by the electron at any worldpoint P_1 we construct the past lightcone corresponding to P_1 (Fig. 4). It intersects the infinite worldline of the electron obviously at a single point P because the tangents to every point on the worldline are all timelike vectors. At P we draw the tangent to the worldline and through P_1 construct the normal P_1Q to this tangent. Let the magnitude of P_1Q be r. According to the definition of a past lightcone the magnitude of PQ should be r/c. *Now the vector of magnitude e/r in the direction PQ represents through its components along the x-, y-, z-axes, the vector potential multiplied by c, and through the component along the t-axis, the scalar potential of the field produced by e at the worldpoint P_1.* This is the essence of the elementary laws formulated by A. Liénard and E. Wiechert.[12]

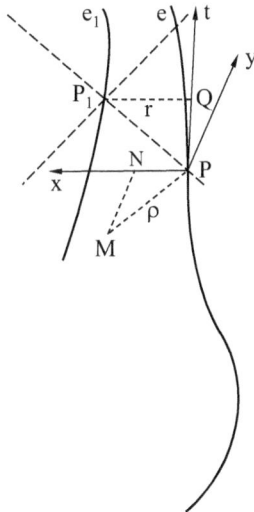

Fig. 4

Then it emerges in the description itself of the field caused by the electron that the division of the field into electric and magnetic forces is a relative one with respect to the specified time axis; most clearly the two forces considered together can be described in some, though not complete, analogy with the wrench in mechanics. I now want to describe *the ponderomotive action of an arbitrarily moving point charge exerted on another arbitrarily moving point charge.* Let us imagine that the worldline of a second pointlike electron of charge e_1 goes through

[12]A. Liénard, "Champ électrique et magnétique produit par une charge concentré en un point et animée d'un mouvement quelconque", L'Éclairage électrique, T. 16, 1898, pp. 5, 53, 106; E. Wiechert, "Elektrodynamische Elementargesetze", Archives Néerlandaiaes des Sciences exactes et naturelles (2), T. 5, 1900, S. 549.

the worldpoint P_1. We define P, Q, r as before, then construct (Fig. 4) the center M of the curvature hyperbola at P, and finally the normal MN from M to an imagined straight line from P parallel to QP_1. We now fix a reference system with its origin at P in the following way: the t-axis in the direction of PQ, the x-axis in the direction of QP_1, the y-axis in the direction of MN, and lastly the direction of the z-axis is determined as being normal to the t-, x-, y-axes. Let the acceleration vector at P be $\ddot{x}, \ddot{y}, \ddot{z}, \ddot{t}$, the velocity vector at P_1 be $\dot{x}_1, \dot{y}_1, \dot{z}_1, \dot{t}_1$. *Now the motive force vector exerted by the first arbitrarily moving electron e on the second arbitrarily moving electron e_1 at P_1 will be*

$$-ee_1(\dot{t}_1 - \frac{\dot{x}_1}{c})\mathfrak{K}$$

where for the components $\mathfrak{K}_x, \mathfrak{K}_y, \mathfrak{K}_z, \mathfrak{K}_t$ of the vector $\mathfrak{K}$ three relations exist:

$$c\mathfrak{K}_t - \mathfrak{K}_x = \frac{1}{r^2}, \quad \mathfrak{K}_y = \frac{\ddot{y}}{c^2 r}, \quad \mathfrak{K}_z = 0$$

and fourthly this vector $\mathfrak{K}$ is normal to the velocity vector at P_1, and this circumstance alone makes it dependent on the latter velocity vector.

If we compare this assertion with the previous formulations[13] of the same elementary law of the ponderomotive action of moving point charges on one another, we are compelled to admit that the relations considered here reveal their inner being in full simplicity only in four dimensions, whereas on a three dimensional space, forced upon us from the beginning, they cast only a very tangled projection.

In mechanics reformed in accordance with the world postulate, the disturbing disharmony between Newtonian mechanics and the modern electrodynamics disappears by itself. In addition, I want to touch on the status of the *Newtonian law of attraction* with respect to this postulate. I will consider two point masses m, m_1, represented by their worldlines, and that m exerts a motive force vector on m_1 exactly as in the case of electrons, except that instead of $-ee_1$ $+mm_1$ should be used. We can now specifically consider the case when the acceleration vector of m is constantly zero, then we may choose t in such a way that m is regarded as at rest, and assume that only m_1 move under the motive force vector which originates from m. If we now modify this specified vector by adding the factor $\dot{t}^{-1} = \sqrt{1 - \frac{v^2}{c^2}}$, which up to magnitudes of the order

---

[13]K. Schwazschild, Nachrichten der k. Gesellschaft der Wissenschaften zu Göttinger, mathematisch-physikalische Klasse, 1903, S. 132; H. A. Lorentz, Enzyklopädie der mathematischen Wissenschaften., V, Art. 14, S. 199.

$1/c^2$ is equal to 1, it can be seen[14] that for the positions x_1, y_1, z_1 of m_1 and their progression in time, we arrive exactly at Kepler's laws, except that instead of the times t_1 the proper times τ_1 of m_1 should be used. On the basis of this simple remark we can then see that the proposed law of attraction associated with the new mechanics is no less well suited to explain the astronomical observations than the Newtonian law of attraction associated with the Newtonian mechanics.

The fundamental equations for the electromagnetic processes in ponderable bodies are entirely in accordance with the world postulate. Actually, as I will show elsewhere, there is no need to abandon the derivation of these equations which is based on ideas of the electron theory as taught by Lorentz.

The validity without exception of the world postulate is, I would think, the true core of an electromagnetic world view which, as Lorentz found it and Einstein further unveiled it, lies downright and completely exposed before us as clear as daylight. With the development of the mathematical consequences of this postulate, sufficient findings of its experimental validity will be arrived at so that even those to whom it seems unsympathetic or painful to abandon the prevailing views become reconciled through the thought of a pre-stabilized harmony between mathematics and physics.

---

[14]H. Minkowski, loc. cit., p. 110.

Appendix B:
On Relativistic Mass

Originally published as an Appendix by the Editor in: A. Einstein, *Relativity*, edited by V. Petkov (Minkowski Institute Press, Montreal 2018)

> *"Mass is a mess"* [1]. Not quite...

During the last three decades physicists have witnessed (or rather endured) "what has probably been the most vigorous campaign ever waged against the concept of relativistic mass"[1] [2].

It seems that campaign had been prompted by Adler's paper "Does mass really depend on velocity, dad?" [3] in which he had even discovered support for his denial of the relativistic mass in Einstein's view on this concept [3, p. 742]:

> Whatever Einstein's precise early views were on the subject, his view in later life appears clear. In a 1948 letter to Lincoln Barnett, he wrote
>
> "It is not good to introduce the concept of the mass $M = m/(1 - v^2/c^2)^{1/2}$ of a body for which no clear definition can be given. It is better to introduce no other mass than the 'rest mass' m. Instead of introducing M, it is better to mention the expression for the momentum and energy of a body in motion."

---

[1]For a detailed account of the controversy over relativistic mass see Chapter 2 of Max Jammer's excellent book *Concepts of Mass in Contemporary Physics and Philosophy* [2].

Unfortunately, Einstein's unclear view of the relativistic mass[2] appears to have provided some encouragement for the campaign against the use of relativistic mass, but the above quote does not in fact demonstrate that "his view in later life appears clear" – Einstein merely expresses his concern and reservation about the definition of M; this becomes evident when it is taken into account that the translation of the above part of Einstein's letter is inaccurate and misleading – compare the translation by Ruschin (there are no such phrases as "introduce no other mass than" and "Instead of introducing M") [7] :

> The German word *daneben* does not mean "instead of," but rather "besides," "in addition to" or "moreover." I would therefore translate the passage:
>
> It is not proper to speak of the mass $M = m/(1 - v^2/c^2)^{1/2}$ of a moving body, because no clear definition can be given for M. It is preferable to restrict oneself to the "rest mass" m. Besides, one may well use the expression for momentum and energy when referring to the inertial behavior of rapidly moving bodies.[3]

Two years after Adler's paper L. B. Okun started a series of publications [8]-[14], which seem to had been the driving force behind the unprecedented campaign against the concept of relativistic mass. In May 1990 *Physics Today* published a number of letters to the Editor with comments on Okun's first article [8] and Okun's replies. W. Rindler's reaction was the sharpest [15]:

> I am disturbed by the harm that Lev Okuns earnest tirade (June 1989 page 31) against the use of relativistic mass ("It is our duty...to stop this process") might do to the teaching of relativity. It might suggest to some who have not thought these matters through that there are unresolved logical difficulties in elementary relativity or that if they use the quantity $m = \gamma m_0$ they commit some physical blunder, whereas in fact this entire ado is about terminology.

Unfortunately, after that exchange "Okuns polemic condemnation" [2, p. 53] even escalated – here are just two examples of his choice of words: "The pedagogical virus of relativistic mass" (from the abstract

---

[2]In his 1905 paper [4] Einstein defined two relativistic masses – longitudinal and transverse masses – but later avoided the entire concept of relativistic mass, including in this book. See [5], [6].

[3]A scan of Einstein's letter in German is included in Okun's article [8].

of a paper [11]) and "The Virus of Relativistic Mass in the Year of Physics" (a title of a paper published in the volume [12]). I think it is truly sad that such a prominent particle physicist did not seem to have even attempted to entertain the possibility that he might have been fundamentally wrong.

While I share the feeling behind Rindler's reaction, I tend to disagree that "this entire ado is about terminology". And papers in support of the relativistic mass do show that the controversy implies more than terminology (see, for example, [16],[17]); here is the conclusion of Bickerstaff and Patsakos' paper [17, p. 66]:

> Thus we conclude by noting that in answering the elementary question of why two different masses are allowed in relativity, one obtains a clearer picture of the subject—a picture that is rooted in mathematics and logic rather than semantics and opinion.

Some physicists have argued that "There is no really good definition of mass" [18]-[21], which might explain the relativistic mass controversy. I tend to disagree with this too. The accepted[4] definition of mass – *the mass of a particle is the measure of the resistance the particle offers to its acceleration* – is both adequate for the concept of mass in relativity and does *indisputably* demonstrate that mass indeed increases with velocity and therefore relativistic mass is an integral part of relativity (complementing proper or rest mass):[5]

- a particle whose velocity increases and approaches the velocity of light *offers an increasing resistance to its acceleration*, that is,

---

[4]Only several examples: "Mass is that property of an object that specifies how much resistance an object exhibits to changes in its velocity" [22]; "mass [is] the resistance of a body to a change of motion" [23]; Mass is "the quantitative or numerical measure of a body's inertia, that is of its resistance to being accelerated" [24]; "We use the term *mass* as a quantitative measure of inertia" [25, p. 9-1]; "Mass...measures how hard we have to push a body to achieve a given acceleration" [26]; "Mass is a quantitative measure of inertia...the greater its mass, the more a body "resists" being accelerated" [27]; "*The qualitative definition of the (inertial) mass of a particle is that it is a numerical measure of the reluctance of the particle to being accelerated*" [28]; "mass is a measure of the inertia of an object" [29]; Mass is defined as the "resistance to acceleration" [30]; "Mass is the measure of the gravitational and inertial properties of matter" [31].

[5]As I think it is exceedingly obvious that there are two masses in relativity (like two times; see below) – rest (or proper) mass and relativistic mass – I do not see any need to comment on the problems (coming from the equivalence of mass and energy) with a *single* concept of mass in relativity ("as an invariant, intrinsic property of an object" [32]).

obviously, its mass (the measure of the resistance the particle offers to its acceleration) increases.[6]

- as in relativity the acceleration of a particle is different in different reference frames, the particle's mass is not an invariant since it is not the same in all frames (only the proper or rest mass, measured in the frame in which the particle is at rest, is an invariant).

These facts make the campaign against the concept of relativistic mass both inexplicable and worrisome. Instead of initiating and stimulating research on the origin of relativistic mass (and on the nature of mass in general) in order to achieve a more profound understanding of this fundamental concept in physics,[7] the relativistic mass is not mentioned at all in many publications[8] (see, for example, the well-known textbook [35]) or, if it is mentioned, it is done to caution the readers[9], that "Most physicists prefer to consider the mass of a particle as fixed" [25, p. 760], that "Most physicists prefer to keep the concept of mass as an invariant, intrinsic property of an object" [32], that "We choose not to use relativistic mass, because it can be a misleading concept" [36] or to warn them [22, p. 1215]:

---

[6]Arguing, effectively, that not the particle's mass but the particle's inertia increases [33] amounts to a rejection of the accepted definition of mass without a valid reason; inertia is the *phenomenon* of offering resistance to acceleration, whereas mass is the *measure* of that resistance.

[7]More research is needed to address the obvious situation: As the *resistance* of a particle to its acceleration depends on the acceleration's direction (the resistance is greater when the acceleration is along the particle's velocity and is becoming infinite as the particle's velocity is approaching the velocity of light), its mass is rather a tensor, not a scalar. In his 1905 paper [4] Einstein defined the two relativistic masses – longitudinal and transverse masses – but later silently abandoned them. With respect to the relativistic masses (longitudinal and transverse) we may witness a repetition of the story with the cosmological constant – initially Einstein used the cosmological constant in his equation linking matter and energy with the spacetime curvature, but later he called it the "biggest blunder of my life;" now cosmologists reintroduced Einstein's cosmological constant. At present time the relativistic mass (let alone the longitudinal and transverse masses) is so out of fashion that even such a prominent relativist as Wolfgang Rindler had to choose the words "confess" and "heuristic" in his letter to the Editor of *Physics Today* [15]: "I will confess to even occasionally using the heuristic concepts of longitudinal mass $\gamma^3 m_0$ and transverse mass γm_0 to predict how a particle will move in a given field of force."

[8]For a list of published works using relativistic mass see [34]. Here I think it is worth mentioning specifically Feynman: "Mass is found to increase with velocity, but appreciable increases require velocities near that of light" [25].

[9]Some authors prefer to take a neutral position: "The use of relativistic mass has its supporters and detractors, some quite strong in their opinions. We will mostly deal with individual particles, so we will sidestep the controversy and use Eq. (37.27) $[\vec{p} = m\vec{v}(1 - v^2/c^2)^{-1/2}]$ as the generalized definition of momentum with m as a constant for each particle, independent of its state of motion" [27, p. 1244]

Watch Out for "Relativistic Mass"

Some older treatments of relativity maintained the conservation of momentum principle at high speeds by using a model in which a particles mass increases with speed. You might still encounter this notion of "relativistic mass" in your outside reading, especially in older books. Be aware that this notion is no longer widely accepted; today, mass is considered as *invariant*, independent of speed. The mass of an object in all frames is considered to be the mass as measured by an observer at rest with respect to the object.

But phrases such as *"prefer* to consider," *"prefer* to keep," *"choose* not to use" (and *"can be"*), *"no longer widely accepted"* and even "older treatments" do not belong to the rigorous language of physics. Physics is not fashion where expressions such as *"prefer to"* and *"choose not to use,"* for example, naturally fit. Physics at its best asks and addresses questions such as:

- Why is the velocity of light the greatest velocity, which cannot be reached by a particle possessing rest mass?

- Why does such a particle offer an *increasing resistance*[10] as its velocity increases and approaches the velocity of light? Or, which is the same question, why does the mass of a particle increase as its velocity increases and approaches the velocity of light?

I think, it is evident from here that the relativistic increase of mass is an *experimental fact* – it is an experimental fact that a particle offers an *increasing resistance* as its velocity increases and approaches the velocity of light and its mass is the measure of that resistance. Therefore the relativistic mass reflects the experimental evidence and we are not free to decide whether or not to use it.

Finally, here I will summarize my response [39] to the objections of Taylor and Wheeler [40] against using the concept of relativistic mass. Here are their objections:

> The concept of 'relativistic mass' is subject to misunderstanding [...]. First, it applies the name mass – belonging to the magnitude of a 4-vector – to a very different concept,

---

[10]In fact, the profound question of the nature of inertia and mass (i.e., the question of the *origin* of the *resistance* a particle offers to its acceleration) has been an open one since Galileo and Newton [37]. The discovery that mass increases with velocity and the controversy over relativistic mass made the need to try to address this open question more urgent.

the time component of a 4-vector. Second, it makes increase of energy of an object with velocity or momentum appear to be connected with some change in internal structure of the object. In reality, the increase of energy with velocity originates not in the object but in the geometric properties of spacetime itself.

It is true that the magnitude of the four-momentum is proportional to the rest mass of a particle:

$$|\vec{p}| = mc \ .$$

The time component of the four-momentum

$$p^0 = \frac{mc}{(1 - v^2/c^2)^{1/2}} = m(v)c$$

is proportional to the relativistic mass $m(v) = m(1 - v^2/c^2)^{-1/2}$. So the rest (proper) mass m is indeed proportional to the magnitude of a four-vector and is an invariant, whereas the relativistic mass $m(v)$ is a component of a four-vector.

However, the situation is precisely the same with respect to proper time and coordinate time. The square of the spacetime distance Δs^2 between two events lying on a timelike worldline is equal to the scalar product $\Delta\vec{x}\cdot\Delta\vec{x}$ of the displacement four-vector $\Delta\vec{x}$ connecting the two events. In other words, the magnitude of the displacement vector is equal to the spacetime distance along the timelike worldline:

$$|\Delta\vec{x}| = \Delta s \ .$$

As $\Delta s = c\Delta\tau$, the magnitude of $\Delta\vec{x}$ is proportional to the proper time $\Delta\tau$ between the two events on the timelike worldline that are connected by the displacement vector:

$$|\Delta\vec{x}| = c\Delta\tau \ .$$

Therefore, the magnitude of the four-vector $\Delta\vec{x}$ is proportional to the proper time $\Delta\tau$.

On the other hand, however, coordinate time is the zeroth (time) component $\Delta x^0 = c\Delta t$ of the displacement four-vector $\Delta\vec{x}$.

So, if we cannot talk about relativistic mass, by the same argument we should talk only about proper time, which is an invariant, and deny the name 'time' to the coordinate time; however, it is the coordinate time that changes relativistically – the experimentally tested time dilation involves precisely coordinate time.

Therefore, proper or rest mass (which is an invariant) and relativistic mass (which is frame-dependent) are exactly like proper time (which is an invariant) and relativistic / coordinate time (which is frame-dependent) [and, to some extent, like proper and relativistic length].

As we saw above, this becomes even more evident from the very definition of mass as the measure of the *resistance* a particle offers to its acceleration or, in the framework of relativity, as the measure of the resistance a particle offers when deviated from its geodesic path. That resistance is different in different reference frames with respect to which the particle moves with different velocities. Therefore the particle mass should also differ in different frames.

It should be stressed that the resistance (and therefore the increased resistance and energy) arises *in* the particle (more precisely, in the particle's worldtube); it does not come from the geometric properties of spacetime. It is spacetime that determines the shape of a geodesic worldline (and the shape of a geodesic worldtube in the case of a spatially extended particle), but it is the particle that *resists* when prevented from "following" a geodesic path, i.e., when the particle's worldtube is *deformed*.

We have proof that the resistance does not originate in the geometry of spacetime – a particle whose worldtube is deformed due to its deviation from its geodesic shape offers the *same* resistance in *both* flat and curved spacetime as the equivalence of inertial and passive gravitational masses shows (for more details see [39, Chap. 9]).

References

1. W. T. Padgett, Problems with the Current Definitions of Mass, *Physics Essays* **3**, 178–182 (1990)

2. M. Jammer, *Concepts of Mass in Contemporary Physics and Philosophy* (Princeton University Press, Princeton 2000) p. 51

3. C. G. Adler, Does mass really depend on velocity, dad? *American Journal of Physics* **55**, 739 (1987)

4. A. Einstein, On The Electrodynamics Of Moving Bodies, *Annalen der Physik* **17** (1905): 891-921, in *The Collected Papers of Albert Einstein*, Volume 2: *The Swiss Years: Writings, 1900-1909* (Princeton University Press, Princeton 1989), pp. 140-171, p. 169

5. E. Hecht, Einstein on mass and energy, *American Journal of Physics* **77**, 799 (2009)

6. E. Hecht, Einstein Never Approved of Relativistic Mass, The Physics Teacher 47, 336 (2009)

7. S. Ruschin, Putting to Rest Mass Misconceptions, *Physics Today*, May 1990, page 15

8. L. B. Okun, The Concept of Mass, *Physics Today* **42**, 31 (1989)

9. . . , , , 1989 . 158, . 3, . 511-530 (L. B. Okun, The Concept of Mass, Usp. Fiz. Nauk. 158, 511 (1989) [Sov. Phys. Usp. 32, 629 (1989)])

10. L. B. Okun, Putting to Rest Mass Misconceptions, *Physics Today* **43**, (1990) pp. 15, 115, 117

11. L. B. Okun, The Concept of Mass in the Einstein Year, 2006, arXiv:hep-ph/0602037

12. L. B. Okun, The Virus of Relativistic Mass in the Year of Physics, in: *Gribov Memorial Volume: Quarks, Hadrons, and Strong Interactions*, Proceedings of the Memorial Workshop Devoted to the 75th Birthday of V. N. Gribov, (World Scientific Publishing, Singapore 2009), pp. 470-473

13. L. B. Okun, Mass versus relativistic and rest masses, *American Journal of Physics* **77**, 430 (2009)

14. L. B. Okun, *Energy and Mass in Relativity Theory*, (World Scientific Publishing, Singapore 2009)

15. W. Rindler, Putting to Rest Mass Misconceptions, *Physics Today*, May 1990, page 13

16. T. R. Sandin, In defense of relativistic mass, *American Journal of Physics* **59**, 1032 (1991)

17. R. P. Bickerstaff and G. Patsakos, Relativistic Generalization of Mass, *European Journal of Physics* **16**, 6368 (1995)

18. E. Hecht, There is no Really Good Definition of Mass, *The Physics Teacher* **44**, 40 (2006)

19. E. Hecht, On Defining Mass, *The Physics Teacher* **49**, 40 (2011)

20. A. Hobson, The Definition of Mass, *The Physics Teacher* **48**, 4 (2010)

21. R. L. Coelho, On the Definition of Mass in Mechanics: Why is it so Difficult? *The Physics Teacher* **50**, 304 (2012)

22. R. A. Serway, J. W. Jewett, Jr., *Physics for Scientists and Engineers with Modern Physics*, 9th ed. (Brooks/Cole, Cengage Learning, Boston 2014) p. 114

23 W. Benenson, J. W. Harris, H. Stocker, H. Lutz (eds), *Handbook of Physics* (Springer, Heidelberg 2002) p. 37

24. S. P. Parker (ed.), *McGraw-Hill Encyclopedia of Physics*, 2nd ed. (McGraw-Hill, New York 1993) p. 762

25. *The Feynman Lectures on Physics, New Millennium Edition*, Vol I (Basic Books, New York 2010) p. 1-2; see also Sections 15-1 (The principle of relativity) and 16-4 (Relativistic mass)

26. M. W. McCall, *Classical Mechanics: From Newton to Einstein: A Modern Introduction*, 2nd ed. (Wiley, Chichester 2011) p. 5

27. H. D. Young, R. A. Freedman, A. L. Ford, *Sears and Zemansky's University Physics with Modern Physics*, 13th ed. (Pearson Education, San Francisco 2012) p. 113

28. D. Gregory, *Classical Mechanics: An Undergraduate Text* (Cambridge University Press, Cambridge 2006) p. 55

29. D. C. Giancoli, *Physics: Principles with Applications*, 7th ed. (Pearson, New York 2014) p. 78

30. W. M. Haynes, D. R. Lide, T. J. Bruno eds, *CRC Handbook of Chemistry and Physics*, 96th ed. (CRC Press, New York, 2015) p. 2-58

31. R. G. Lerner, G. L. Trigg, *Encyclopedia of Physics*, 2nd ed. (VCH Publishers, New York 1991) p. 703

32. S. T. Thornton, A. Rex, *Modern Physics for Scientists and Engineers*, 4th ed. (Brooks/Cole, Cengage Learning, Boston 2013) p. 61

33. J. Roche, What is mass? *European Journal of Physics* **26** (2005) pp. 1-18

34. G. Oas, On the Use of Relativistic Mass in Various Published Works, 2005, arXiv:physics/0504110 [physics.ed-ph]; see also G. Oas, On the Abuse and Use of Relativistic Mass, 2005, arXiv:physics/0504110 [physics.ed-ph]

35. J. Walker, D. Halliday, R. Resnick, *Fundamentals of Physics Extended*, 10th ed. (John Wiley, New York 2014)

36. K. S. Krane, *Modern Physics* 3rd ed. (Wiley, Chichester 2012) p. 51

37. Newton explicitly defined inertia as the *resistance* a body offers to a change of its velocity (boldface added to "resisting" – V.P.): *"Inherent force of matter is the power of **resisting** by which every body, as far as it is able, perseveres in its state either of resting or of moving uniformly straight forward"* [38]

38. I. Newton, *The Principia: Mathematical Principles of Natural Philosophy*, A new translation by I. Bernard Cohen and Anne Whitman assisted by Julia Budenz (University of California Press, Berkeley 1999) p. 404

39. V. Petkov, *Relativity and the Nature of Spacetime*, 2nd ed. (Springer, Heidelberg 2009) p. 115

40. E. F. Taylor, J. A. Wheeler: *Spacetime Physics: Introduction to Special Relativity*, 2nd ed. (Freeman, New York 1992) pp. 250-251

114

Appendix C: On inertial forces, inertial energy and the origin of inertia

Vesselin Petkov

Institute for Foundational Studies 'Hermann Minkowski'
Montreal, Quebec, Canada
http://minkowskiinstitute.org/
vpetkov@minkowskiinstitute.org

Abstract

As a result of the open question of inertia the status of inertial forces has been a difficult subject in physics with implications for the proper understanding of the force of weight in general relativity where gravity is not a force, but a manifestation of the spacetime curvature. The purpose of this paper is fourfold. First, to state explicitly when the inertial forces are fictitious and when real. Second, to provide a virtually self-evident derivation, which demonstrates that kinetic energy is in fact inertial energy – the energy equal to the work done by inertial forces. Third, to stress that weight, which has been traditionally regarded as a gravitational force, is an inertial force in general relativity. Fourth, to outline what relativity implies about the origin of inertia.

Keywords Inertial forces, inertial energy, kinetic energy, geodesic hypothesis, origin of inertia, four-dimensional stress

The best way to approach the issue of inertial forces is by recalling the definition of mass which has been adopted since Newton – mass is the measure of the *resistance* a particle offers to its acceleration. It is this resistance, commonly called inertia, which *experimentally* distinguishes accelerated from inertial motion. Due to the fact that the presence or the absence of a particle's resistance to its motion is *absolute* or frame-independent, both accelerated and inertial motion are absolute or frame independent. An accelerating particle's resistance allows its state of absolute acceleration to be detected in any reference frame. Similarly, the absence of resistance to a particle's motion makes the detection of the particle's inertial motion possible in any reference frame. For this reason a particle's resistance to its motion is a necessary and sufficient condition for it to be in a state of absolute acceleration caused by some interaction. Conversely, if a particle does not resist its motion, it is a free particle, which is not subject to any interactions and moves by inertia.

In his spacetime formulation of special relativity Minkowski provided rigorous criteria for inertial and accelerated motion [1] – a free particle, which moves by inertia, is a straight timelike worldline in Minkowski spacetime, whereas the timelike worldline of an accelerating particle is curved. These criteria show that in spacetime the absoluteness of accelerated and inertial motion become even more understandable – the straightness of a timelike worldline (representing inertial motion) and the curvature or rather the *deformation* of a timelike worldline (representing accelerated motion) are absolute (frame independent) properties of worldlines. Therefore it is the *deformation* of the worldline of an accelerating particle that makes the particle's acceleration absolute[1]. In such a way Minkowski's spacetime representation of special relativity unequivocally supported Newton's view of absolute acceleration and disproved Mach's arguments that acceleration, like velocity, is also relative[2]. Also, the proper relativistic understanding of the absoluteness of acceleration demonstrates that absolute acceleration merely reflects the deformation of an accelerating particle's worldline and does not imply some absolute space with respect to which the particle accelerates.

Now the distinction between fictitious and real inertial forces can be demonstrated by a simple example involving an accelerating elevator. Let two elevators I and N be at relative rest far away from gravitating masses and let a ball be floating in the middle of N. At a given moment N starts to accelerate translationally and observers inside it see that the ball starts to fall (accelerate) towards the elevator's floor. The apparent

---

[1] Acceleration as a deformation of a geodesic worldline is absolute in both special and general relativity. There is a second acceleration in general relativity which, however, is not related to a deformation of the geodesic worldline of an apparently accelerating particle and it is *non-resistant* since it is relative or apparent – the absence of parallel, or rather congruent, geodesic worldlines in non-Euclidean spacetime leads to geodesic deviation, which manifests itself as a relative acceleration in general relativity.

[2] Mach argued that one could not say whether or not a *single* particle in the Universe accelerates. By contrast, that situation in spacetime is crystal clear – the worldline of a single particle in the Universe is either straight or deformed, which means that the particle is either moving by inertia or with an acceleration.

accelerated motion of the ball can be formally regarded as caused by a force. However, due to the fact that the presence of acceleration is absolute (frame independent), observers in both the inertial elevator I and the non-inertial elevator N agree that it is N that accelerates, not the ball; it is N's floor that *in reality* approaches the ball. And there is no relativity here since an accelerometer attached to the ball detects no acceleration. This is best seen in the spacetime diagram depicted in Fig. 6.2. For this reason the force formally introduced to explain the ball's apparent acceleration in N is not a real force; it is a fictitious inertial force, which can be imagined when the inertial motion of the ball is described in the non-inertial elevator N. The situation changes when N's floor reaches the ball and starts to accelerate it. The ball *resists* the change in its inertial state and exerts a real translational inertial force back on the floor.

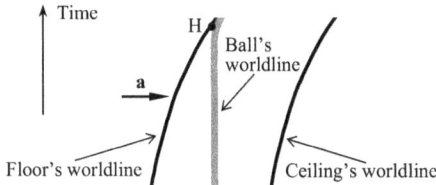

Figure 6.2: A starting to accelerate (with acceleration **a**) elevator N is represented by the worldlines of its ceiling and floor. A ball which was floating in the middle of N before N started to accelerate is represented by its worldline. The ball's worldline is straight, which means that it does not accelerate and therefore moves by inertia. It is the floor's worldline that converges towards the straight worldline of the ball, or in the ordinary three-dimensional language it is the floor that *in reality* approaches the ball. At event H the floor hits the ball and starts to accelerate it, the ball's worldline is deformed and the ball resists its acceleration by acting back on the floor with a *real* inertial force.

This distinction between fictitious and real inertial forces in the case of translational acceleration holds also for the cases of rotational non-inertial motion. Translational, centrifugal, and Coriolis inertial forces are fictitious when a *free* particle (moving non-resistantly by inertia) *only appears to be accelerated by a fictitious inertial force* when described in a non-inertial reference frame. When the particle is compelled to move with the non-inertial frame's acceleration, it starts to resist the change in its inertial motion and exerts a *real inertial force* on the mover that accelerates it.

What also might contribute to a better understanding of the status of inertial forces is the fact that *real inertial forces do work*, which implies that kinetic energy is rather *inertial energy*. In the above case the deformation on N's floor (resulting from the collision of the ball and the floor) is caused by the real inertial force with which the ball resists its acceleration. Therefore the work done by the ball's inertial force, which is equal to its inertial energy, converts into a deformation energy. So far inertial energy has been called kinetic energy. But such a name does not reveal the true nature of the ball's energy responsible for the deformation on the N's floor – the ball's inertia, i.e. its resistance to the change in its inertial state.

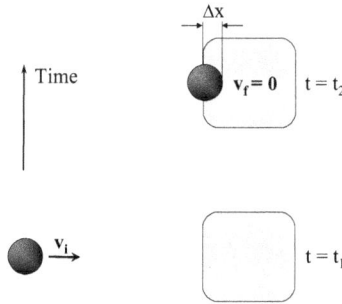

Figure 6.3: A massive plastic block is deformed when hit by a ball moving by inertia. Traditionally, it is stated that the ball's kinetic energy converts into a deformation energy. However, a deep physical explanation reveals that the ball's energy is inertial energy since the deformation is caused by the work done by the real inertial force with which the ball resists its deceleration

The qualitative argument that kinetic energy is actually inertial energy has a straightforward quantitative counterpart. That inertial energy – the work done by inertial forces – is equal to kinetic energy is easily demonstrated by an example depicted in Fig. 6.3. At moment $t = t_1$ a ball travels at constant "initial" velocity v_i towards a huge block of some plastic material; we can imagine that the block is mounted on the steep slope of a mountain. Immediately after that the ball hits the block, deforms it and is decelerated. At moment $t = t_2$ the block stops the ball, that is, the ball's final velocity at t_2 is $v_f = 0$ (the block's mass is effectively equal to the Earth's mass, which ensures that $v_f = 0$). According to the standard explanation it is the ball's kinetic energy $E_k = (1/2)mv_i^2$ which transforms into a deformation energy. But a proper physical explanation demonstrates that the energy of the ball, which is transformed into deformation energy, is its inertial energy E_i, because the ball resists its deceleration a and it is the work $W = F\Delta x$ (equal to E_i) done by the *inertial* force $F = ma$ that is responsible for the deformation of the plastic material.

Using the relation between v_i, v_f, a and the distance Δx in the case of deceleration

$$v_f^2 = v_i^2 - 2a\Delta x$$

and taking into account that $v_f = 0$ we find

$$a = \frac{v_i^2}{2\Delta x}.$$

Then for the ball's inertial energy E_i we have

$$E_i = W = F\Delta x = ma\Delta x = \frac{1}{2}mv_i^2.$$

Therefore the inertial energy of the ball is indeed equal to what has been descriptively (lacking physical depth) called kinetic energy.

Minkowski's distinction between inertial and non-inertial motion has been generalized in general relativity by the *geodesic hypothesis* – a *free* particle is a timelike *geodesic* worldline in curved spacetime. The geodesic hypothesis is "a natural generalization of Newton's first law" [2], that is, "a mere extension of Galileo's law of inertia to curved space-time" [3]. Therefore *in general relativity a particle, whose worldline is geodesic, moves by inertia.* A particle, whose worldline is deformed (that is, is not a geodesic), resists its deviation from its geodesic (i.e. inertial) path in spacetime, and exerts a real inertial force on the obstacle that deforms the particle's worldline.

The geodesic hypothesis has been proved by the experimental fact that particles falling towards the Earth's surface *do not resist* their fall[3] (a falling accelerometer reads zero acceleration or rather zero resistance), which means that they indeed move by inertia while falling. If a falling particle were subject to a gravitational force it would *resist* its fall (its apparent acceleration) because by Newton's second law *a force is only needed to overcome an accelerating particle's resistance.* When a falling particle hits the ground it is prevented from moving by inertia and resists its resulting absolute acceleration[4] (while being at rest on the ground,

---

[3]In the case of small particles tidal effects can be safely ignored. But even if tidal effects are taken into account, the tidal friction has nothing to do with the resistance a particle which is subject to a force offers to its acceleration. The tidal effects are merely a manifestation of the spacetime curvature.

[4]It follows from here that, like the inertial mass, the passive gravitational mass can be also defined as the measure of the resistance a particle offers to its acceleration, which sheds additional light on the equivalence of the two masses [4].

the particle's worldline is *deformed* and by the generalized Minkowski criterion the particle is subject to absolute acceleration). Two things are now evident – (i) the particle's weight is the resistance force, which it exerts on the ground, and (ii) that resistance force is *inertial* since it is the force with which the particle *resists* the change in its inertial motion (its fall).

The relativistic explanation of the absoluteness of acceleration as a *deformation* of the worldline or rather the worldtube of an accelerating particle provides an unanticipated insight into the origin of inertia. The resistance an accelerating particle offers to its acceleration (i.e. to the deformation of its worldtube) can be regarded as a manifestation of a static restoring force caused by a four-dimensional stress that arises in the deformed worldtube[5] of the accelerating particle (like the restoring force arising when an ordinary three-dimensional rod is deformed) [4, 5]. Therefore relativity implies that inertia has a *local* origin – it is the accelerating particle *itself* that resists the deformation of its worldtube. This means that inertia is not a non-local phenomenon that is caused by the distant masses as Mach argued. It might be tempting to say that what determines the shape of a free particle's geodesic worldtube (which when deformed resists its deformation) are all the masses in the Universe in line with Mach's view of inertia as caused by the distant masses. However, such a temptation would be misleading since in curved spacetime it is the nearby mass that is essentially responsible for the shape of the geodesics in its vicinity. The shape of the geodesic worldline of a particle falling towards the Earth, for example, is predominantly

---

[5]This explanation of the origin of inertia presupposes that the worldtubes of particles are real four-dimensional objects, which is a direct consequence of Minkowski's view of special relativity as a theory of an absolute four-dimensional world and particularly of his explanation of length contraction. In Minkowski's explanation of this relativistic effect the instantaneous three-dimensional spaces of two observers in relative motion intersect the worldtube (Minkowski called it the world strip) of a body at different angles and the two resulting three-dimensional cross-sections have different lengths. It is obvious that these cross-sections of the worldtube of the *same* three-dimensional body would be impossible if its worldtube were not real, i.e. if it were a mere geometrical abstraction [1] (see also [4, Chap. 5], [5]). In fact, the reality of the worldtubes of physical objects and the four-dimensionality of the world itself (at least at the macro scale) follows from the experimental evidence supporting the relativity principle as first realized by Minkowski. He noticed that as observers in relative motion, according to the relativity principle, have *different times* (the realization of which led Einstein to the special theory of relativity), it follows that the observers also have *different spaces* – "Hereafter we would then have in the world no more *the* space, but an infinite number of spaces analogously as there is an infinite number of planes in three-dimensional space. Three-dimensional geometry becomes a chapter in four-dimensional physics." [1]. Obviously, many three-dimensional spaces imply a four-dimensional world.

determined by the Earth's mass and the distant masses have practically zero contribution.

References

[1] H. Minkowski, Space and Time. New translation in: H. Minkowski, *Space and Time: Minkowski's Papers on Relativity* (Minkowski Institute Press, Montreal 2012) pp. 111–125 (http://minkowskiinstitute.org/mip/); included in this book as Appendix A

[2] J. L. Synge, *Relativity: the general theory.* (Nord-Holand, Amsterdam 1960) p 110

[3] W. Rindler, *Relativity: Special, General, and Cosmological* (Oxford University Press, Oxford 2001) p 178

[4] V. Petkov, *Relativity and the Nature of Spacetime*, 2nd ed. (Springer, Heidelberg 2009), Chap. 9

[5] V. Petkov, On the Reality of Minkowski Space, *Foundations of Physics* **37** (2007) 1499–1502

Appendix D: Can Gravity be Quantized?

Vesselin Petkov

Institute for Foundational Studies 'Hermann Minkowski'
Montreal, Quebec, Canada
http://minkowskiinstitute.org/
vpetkov@minkowskiinstitute.org

Abstract

It may turn out that we have been stubbornly ignoring a crucial message coming from the unsuccessful attempts to create a theory of quantum gravity – that gravity is not an interaction. This option does not look so shocking when gravity is consistently and rigorously regarded as a manifestation of the non-Euclidean geometry of spacetime. Then it becomes evident that general relativity does imply that gravitational phenomena are not caused by gravitational interaction. The geodesic hypothesis in general relativity and particularly the experimental evidence that confirmed it indicate that gravity is not a physical interaction since particles which appear to interact gravitationally are actually free particles whose motion is inertial (i.e. interaction-free). This situation has implications for two research programs – quantum gravity and detection of gravitational waves. First, the real open question in gravitational physics appears to be how matter curves spacetime, not how to quantize the apparent gravitational interaction. Second, the search for gravitational waves should explicitly take into account the geodesic hypothesis according to which orbiting astrophysical bodies (modelled by point masses) do not radiate gravitational energy since their worldlines are geodesics representing inertial (energy-loss-free) motion.

6.1 Introduction

Since the advent of general relativity and quantum mechanics their unification has been the ultimate goal of theoretical physics. So far, however, the different approaches aimed at creating a theory of quantum gravity [1] have been unsuccessful. It seems a possible reason for this – that gravity might not be an interaction – has never been consistently examined. What also warrants such an examination is that an experimental fact – falling bodies *do not resist* their apparent acceleration – turns out to be crucial for determining the true nature of gravitational phenomena, but has been effectively neglected so far. Taking it into account, however, makes it possible to refine not only the quantum gravity research (by recognizing that the genuine open question in gravitational physics is how matter determines the geometry of spacetime, not how to quantize what has the appearance of gravitational interaction) but also to fine-tune the search for gravitational waves by showing that astrophysical bodies, modelled by point masses whose worldlines are geodesics (representing inertial or energy-loss-free motion), do not give rise to radiation of gravitational energy.

As too much is at stake in terms of both the number of physicists working on quantum gravity and on detection of gravitational waves, and the funds being invested in these worldwide efforts, even the heretical option of not taking gravity for granted should be thoroughly analyzed. It should be specifically stressed, however, that such an analysis will certainly require extra effort from relativists who are more accustomed to solving technical problems than to examining the physical foundation of general relativity which may involve no calculations. Such an analysis is well worth the effort since it ensures that what is calculated is indeed in the proper framework of general relativity and is not smuggled into it to twist it until it yields some features that resemble gravitational interaction.

The standard interpretation of general relativity takes it as virtually unquestionable that gravitational phenomena result from gravitational interaction. However, the status of gravitational interaction in general relativity is far from self-evident and its clarification needs a careful analysis of both the mathematical formalism and the logical structure of the theory and the existing experimental evidence.

Taken according to its logical structure general relativity demonstrates that what is traditionally called gravitational interaction is dramatically different from the other three fundamental interactions, successfully described by the Standard Model, and is nothing more than a mere manifestation of the curvature of spacetime. Unlike the electromagnetic interaction, for example, which is mediated by the electromagnetic field and force, the observed apparent gravitational interaction is not caused by a *physical* gravitational field and a gravitational force. By the geodesic hypothesis in general relativity, the assumption that the worldline of a free particle is a timelike *geodesic* in spacetime is "a natural generalization of Newton's first law" [2, p. 110], that is, "a mere extension of Galileo's law of inertia to curved spacetime" [3, p. 178]. This means that *in general relativity a particle, whose worldline is geodesic, is a free particle which moves by inertia.*

Indeed, two particles that seem to be subject to gravitational forces in reality *move by inertia* according to general relativity since their worldlines are timelike geodesics in spacetime curved by the particles' masses. The acceleration of the particles towards each other is *relative* and is caused not by gravitational forces, but by geodesic deviation, which reflects the fact that there are no straight worldlines in curved spacetime. In general relativity the planets, for example, are free bodies which move by inertia and as such do not interact in any way with the Sun because *inertial motion does not imply any interaction.* The planets' worldlines are geodesics[1], which due to the curvature of spacetime caused by the Sun's mass are helixes around the worldline of the Sun (which means that the planets move *by inertia* while orbiting the Sun).

Therefore, what general relativity *itself* tells us about the world is that the apparent gravitational interaction is not a physical interaction in a sense that two particles, which *appear* to interact gravitationally, are free particles since they move by inertia. This readily, but counterintuitively explains the unsuccessful attempts to create a theory of quantum gravity – it is impossible to quantize what we regard as gravitational interaction since it simply does not exist according to what the logical structure of general relativity *itself* implies (without importing features to general relativity whose sole justification is the *belief* that gravitational phenomena are caused by gravitational interaction).

Two main reasons have been hampering the proper understanding of gravitational phenomena. The first reason, discussed in Sect. 6.2, is that the profound consequences of the geodesic hypothesis for the nature of gravitational interaction have not been fully realized mostly due to the adopted definition of a free particle in general relativity, which literally posits that otherwise free particles are still subject to gravitational interaction – an assumption that does not follow from the theory itself. Sect. 6.3 examines the second reason – that since the advent of general relativity there have been persistent attempts to squeeze general relativity and ultimately Nature into the present understanding that gravitational energy and momentum (as energy and momentum of gravitational interaction and field) are part of gravitational phenomena.

---

[1]Only the center of mass of a spatially extended body is a geodesic worldline.

6.2 General relativity implies that there is no gravitational interaction

The often given definition of a free particle in general relativity – a particle is "free from any influences other than the curvature of spacetime" [5] – effectively *postulates* the existence of gravitational interaction by almost explicitly asserting that the influence of the spacetime curvature on the shape of a free particle's worldline constitutes gravitational interaction.

However, if carefully analyzed, the fact that particles' masses curve spacetime, which in turn changes the shape of the worldlines of those particles, does not imply that the particles interact gravitationally. There are two reasons for that. First, the shape of the geodesic worldlines of free particles is determined by the curvature of spacetime *alone* which itself may not be necessarily induced by the particles' masses. This is best seen from the fact that general relativity shows both that spacetime is curved by the presence of matter, and that a matter-free spacetime can be *intrinsically* curved. The latter option follows from the de Sitter solution [4] of Einstein's equations. Two test particles in the de Sitter universe only appear to interact gravitationally but in fact their interaction-like behaviour is caused by the curvature of their geodesic worldlines, which is determined by the constant positive *intrinsic* curvature of the de Sitter spacetime. The fact that there are no straight geodesic worldlines in non-Euclidean spacetime (which gives rise to geodesic deviation) manifests itself in the relative acceleration of the test particles towards each other which creates the impression that the particles interact gravitationally (*test* particles' masses are assumed to be negligible in order not to affect the geometry of spacetime).

Second, the experimental fact that particles of different masses fall towards the Earth with the *same* acceleration in full agreement with general relativity's *"a geodesic is particle-independent"* [3, p. 178], ultimately means that the shape of the geodesic worldline of a free particle in spacetime curved by the presence of matter is determined by the spacetime geometry *alone* and *not by the matter*. This is clearly seen when the central point of general relativity – the mass-energy of a body changes the geometry of spacetime around itself – is explicitly taken into account. The very meaning of changing the geometry of *empty* spacetime by a body is that *the geodesics of the new spacetime geometry are set*. This is so because what essentially determines the type of spacetime geometry is the corresponding version of Euclid's fifth postulate, which is expressed in terms of the geodesic worldlines of the spacetime geometry. Hence a geodesic is particle-independent because a geodesic is a feature of the spacetime geometry *itself.* The fact that the worldline of a free particle is influenced by the curvature of spacetime produced by a body does not constitute gravitational interaction with the body since the shape of the free particle's worldline is not changed by the body's mass-energy – the body curves *solely* spacetime, regardless of whether or not spacetime is empty, because *no additional energy is spent* for curving the geodesic worldline of the free particle (or in three-dimensional language – no additional energy is spent for making the particle orbit the body or fall onto it). In short, the mass-energy of a body changes the geometry of spacetime no matter whether or not there are any particles in the body's vicinity, and the shape of free particles' worldlines reflects the spacetime curvature no matter whether it is intrinsic or induced by a body's mass-energy.

Another indication that the shape of a geodesic worldline is set by the geometry of spacetime *alone*, irrespective of the origin of the spacetime geometry itself, is the fact that the spacetime curvature created by the mass-energy of a free body determines not only the shape of the worldline of a second nearby body, but also the shape of the first body's *own* geodesic worldline. This is best seen in the binary star systems where two stars orbit the common center of gravity which lies between them. As each star curves spacetime regardless of whether or not the other star is there, the resulting spacetime curvature is a superposition of the curvatures produced by the two stars. In the so curved spacetime the geodesic worldlines of the starts' centers are helixes around the worldline of the common center of gravity (curvature). So the worldline of each star's center is determined by the spacetime curvature induced by *both* stars; by contrast, if one of the stars were a particle with much smaller mass compared to the mass of the other star, the particle's geodesic worldline would be a helix around the geodesic worldline of the center of the other star. If it is assumed that the shape of the stars' geodesic worldlines were not determined *solely* by the spacetime geometry, but were a result of gravitational interaction between the two stars – that is, if it were the mass-energy of each star that determined through the spacetime curvature the shape of the other star's worldline – it would be then a mystery how each star would interact gravitationally with itself, i.e. how the mass-energy of each star would also determine the shape of its own worldline. This problem does not arise when it is recognized that the shapes of the stars' worldlines merely reflect the spacetime curvature regardless of whether or not it is intrinsic or created by the stars' masses. Simply, the mass-energy of each star individually and independently from the other star changes the spacetime geometry, and the shapes of the stars' geodesics reflect the resultant spacetime curvature. (Of course, this problem does not arise in the Newtonian gravitational theory since it regards gravitational phenomena as caused by a force, not as a manifestation of spacetime curvature.)

The essential role of inertial motion in general relativity follows from the basic fact that the existence of geodesics is a feature of curved spacetime *itself* just like the existence of straight worldlines is a feature of flat spacetime. Straight worldlines represent the inertial motion of free particles of *any* mass in flat spacetime and the straightness of their worldlines is regarded as naturally reflecting the spacetime geometry. Analogously, geodesics in curved spacetime represent free particles of *any* mass that move by inertia. The shape of the geodesics also reflects the spacetime geometry and is not an indication of some interaction exactly like the shape of the straight worldlines in flat spacetime is not an indication of any interaction. The equal status of geodesics in flat and curved spacetimes is encoded in the fall of different masses with the same acceleration. By the geodesic hypothesis, their fall is inertial and indeed the motion of falling particles is unsurprisingly similar to motion by inertia in the absence of gravity – particles that move by inertia do so irrespective of their masses.

That a geodesic worldline in curved spacetime represents an *unconditionally* free particle becomes clearer from a closer examination of the geodesic hypothesis itself and particularly from the experimental evidence which proved it.

Newton's first law of motion (i.e. Galileo's law of inertia) describes the motion of a free particle that is not subject to any interactions. Such a particle moves by inertia, which means that it *offers no resistance* to its motion with constant velocity. If a particle is subject to some interaction, which prevents it from maintaining its inertial motion, the particle *resists* the forced change of its velocity, i.e. the particle resists its acceleration. The particle's reaction and its resistance to the interaction is captured in Newton's third and second laws of motion. The third law reflects the fact that when a free particle is subject to some action it offers an equal and opposite reaction by *resisting* the action. The profound meaning of Newton's second law is that *a force is only needed to overcome the resistance the particle offers to its acceleration.*

It is the intrinsic feature of a particle to move *non-resistantly* by inertia when its motion is not disturbed by *any* influences that constitutes an objective criterion for a free particle. That is, *non-resistant* motion is a necessary and sufficient condition for a particle to be *free*. *A particle is subject to some interaction only if it resists its motion.*

Galileo's and Newton's law of inertia was first generalized in special relativity by Minkowski who realized that a free particle, which moves by inertia, is a straight timelike worldline in Minkowski spacetime [6]. By contrast, the worldline of an accelerating particle is curved, i.e. deformed. Had this generalization of the law of inertia been carefully analyzed, two immediate consequences would have been realized. First, the experimental fact that acceleration is *absolute*, because it is detectable due to the resistance an accelerating particle offers to its acceleration, finds an unexpected but deep explanation in Minkowski's spacetime formulation of special relativity. The acceleration of a particle is absolute not because the particle accelerates with respect to some absolute space, but because its worldline is curved and therefore *deformed*, which is an absolute geometric property that corresponds to the absolute physical property of the particle's resistance to its acceleration. Second, the resistance an accelerating particle offers to its acceleration can be also given an unforeseen explanation – as the worldline or rather the world-tube of an accelerating particle is curved, the particle's resistance to its acceleration (i.e. the particle's inertia) can be viewed as originating from a four-dimensional stress which arises in the *deformed* worldtube of the particle[2] [7, Chap. 9].

Based on Minkowski's rigorous definition of a free particle in special relativity, the above criterion for a free particle can be made even more precise – in three-dimensional language, a free particle does not resist its motion, whereas in four-dimensional (spacetime) language a free particle is a timelike worldtube, which is *not deformed*. And indeed, in Minkowski spacetime straight worldtubes are not distorted, which explains why a free particle, represented by a straight worldtube, offers no resistance to its free or inertial motion. This criterion provides further justification for the geodesic hypothesis in general relativity by clarifying that a timelike geodesic worldtube in curved spacetime, which represents a free particle, is naturally curved due to the spacetime curvature, but is not deformed[3].

The generalized Minkowski definition of a free particle in spacetime (no matter flat or curved) – a free particle is a *non-deformed* worldtube (straight in flat spacetime and geodesic in curved spacetime) – indicates

---

[2]This explanation of inertia implies that the worldtubes of particles are real four-dimensional objects completely in line with Minkowski's view of special relativity as a theory of an absolute four-dimensional world and particularly with his explanation of length contraction, which would be impossible if the worldtube of a relativistically contracted body were not real, i.e. if it were a mere geometrical abstraction [6] (see also [7, 8]).

[3]Rigorously speaking, this is true only for a small (test) particle. Tidal stresses, caused by geodesic deviation, give rise to some deformation but that is not caused by a gravitational *force*.

that a geodesic worldline does represent an unconditionally free particle in general relativity. Indeed, no interaction is behind the fact that the worldtube of a free particle in flat spacetime is straight and the same is true for a free particle in curved spacetime – no interaction is responsible for *the curved but not deformed* geodesic worldtube of a free particle there (in agreement with the fact that a geodesic worldline is the analog of a straight worldline in curved spacetime).

What is crucial for testing both the geodesic hypothesis and the generalized definition of a free particle in spacetime and for determining the true nature of gravitational phenomena is the experimental fact that particles falling towards the Earth's surface *offer no resistance to their fall.* This essential experimental evidence has been virtually neglected so far, which is rather inexplicable especially given that Einstein regarded the realization of this fact – that "if a person falls freely he will not feel his own weight" – as the "happiest thought" of his life which put him on the path towards general relativity [9].

This experimental fact unambiguously confirms the geodesic hypothesis because free falling particles, whose worldtubes are geodesics, do not resist their fall (i.e. their apparent acceleration) which means that they move by inertia and therefore no gravitational force is causing their fall. It should be particularly stressed that a gravitational force would be required to accelerate particles downwards *only if* the particles *resisted* their acceleration, because *only then* a gravitational force would be needed to *overcome* that resistance.

Thus, the experimental evidence of non-resistant fall of particles is the definite proof of the central assumption of general relativity – that no gravitational force is causing the gravitational phenomena. This experimental evidence is crucial since it rules out any alternative theories of gravity and any attempts to quantize gravity (by proposing alternative representations of general relativity aimed at making it amenable to quantization) that regard gravity as a *physical* field which gives rise to a gravitational *force* since they would contradict the experimental evidence.

The non-resistant fall of particles also confirms the generalized definition of a free particle since their geodesic worldtubes are naturally curved (due to the spacetime curvature) but are *not deformed.* A falling accelerometer, for example, reads zero acceleration (in an apparent contradiction with the observed acceleration of the accelerometer while falling), which is adequately explained when it is taken into account that what an accelerometer measures is the *resistance* it offers to its acceleration. The zero reading of the falling accelerometer proves that it offers no resistance to its fall and demonstrates that it moves by inertia and therefore its acceleration is not absolute (not resulting from a *deformation* of its worldtube); it is *relative* due to its naturally curved, but *not deformed* worldtube (that is, the accelerometer's relative acceleration is caused by geodesic deviation which itself is a manifestation of the fact that the geodesic worldtube of the accelerometer and the worldline of the Earth's center converge towards each other).

The accelerometer does not resist its fall because its absolute acceleration is zero according to general relativity ($a^\mu = d^2x^\mu/d\tau^2 + \Gamma^\mu_{\alpha\beta}(dx^\alpha/d\tau)(dx^\beta/d\tau) = 0$), which reflects the fact that its worldtube is geodesic and is therefore not deformed[4]. When the accelerometer is at rest on the Earth's surface its worldtube is not geodesic, which by the geodesic hypothesis means that the accelerometer does not move by inertia and therefore should resist its being prevented from maintaining its inertial motion, i.e. the accelerometer should resist its state of rest on the Earth's surface. Before the advent of general relativity that resistance force had been called gravitational force or the accelerometer's weight. As implied by the geodesic hypothesis the accelerometer's weight is the inertial force, which arises when the accelerometer is prevented from moving by inertia while falling. This is also seen from the fact that the accelerometer's worldtube is *deformed* (not geodesic) – the four-dimensional stress in the deformed worldtube gives rise to a static restoring force that manifest itself as the resistance (inertial) force with which the accelerometer opposes its deviation from its geodesic path in spacetime. The concept of inertia in Minkowski's spacetime formulation of special relativity sheds more light on the physical meaning of the equivalence of inertial and (passive) gravitational masses and forces. They are all inertial and originate from the four-dimensional stress arising in the deformed worldtubes of non-inertial particles (accelerating or being prevented from falling) [7, Ch. 9]. So in the framework of relativity the definition of mass as the measure of the resistance a body offers to its acceleration (i.e. to the deformation of its worldtube) becomes

---

[4]Had Minkowski lived longer he might have discovered general relativity (surely under another name) before Einstein. Minkowski would have almost certainly noticed that inertia could be regarded as arising from the four-dimensional stress in the deformed worldtube of an accelerating particle and therefore inertia would turn out to be another manifestation (along with length contraction as correctly explained by him) of the four-dimensionality of the absolute world of his spacetime formulation of special relativity. Then the experimental fact that a falling particle accelerates (which means that its worldtube is curved), but offers no resistance to its acceleration (which means that its worldtube is not deformed) can be explained only if the worldtube of a falling particle is both curved and not deformed, which is impossible in the flat Minkowski spacetime where a curved worldtube is always deformed. Such a worldtube can exist only in a non-Euclidean spacetime whose geodesics are naturally curved due to the spacetime curvature, but are not deformed.

even more understandable.

6.3 There is no gravitational energy in general relativity

The second main reason, which has been hampering the proper understanding of gravitational phenomena, is the issue of gravitational energy and momentum.

Einstein made the gigantic step towards the profound understanding of gravity as spacetime curvature but even he was unable to accept all implications of the revolutionary view of gravitational phenomena. It was he who first tried to insert the concepts of gravitational energy and momentum forcefully into general relativity in order to ensure that gravity can still be regarded as some interaction despite that the mathematical formalism of general relativity itself refused to yield a proper (tensorial) expression for gravitational energy and momentum. This refusal is fully consistent with the status of gravity as non-Euclidean spacetime geometry (*not a force*) in general relativity. The non-existence of gravitational force implies the non-existence of gravitational energy as well since gravitational energy *presupposes* gravitational force (gravitational energy = work due to gravity = gravitational force times distance).

Although the mathematical formalism and the logical structure of general relativity imply that gravitational phenomena are not caused by gravitational interaction, which entails that there are no gravitational energy and momentum in Nature, most relativists regard gravitational energy as a necessary element of the description of gravitational phenomena. This position is based not only on the view, which literally postulates the existence of gravitational interaction and therefore of gravitational energy and momentum, but also on two generally accepted views.

First, the nonlinearity of Einstein's equations has been interpreted to support the assumption that like the electromagnetic field, the gravitational field also carries energy and momentum. However, unlike Maxwell's equations, which are linear because the electromagnetic field itself does not have electric charge and does not contribute to its own source, the gravitational field must contribute to its own source if it carries energy and momentum since in general relativity any energy is a source of gravity. This would be consistent with the fact that Einstein's equations are nonlinear – the nonlinearity would represent the effect of gravitation on itself. However, this interpretation of Einstein's equations barely hides the major problem of the standard interpretation of generally relativity that there exists gravitational interaction and therefore gravitational field, which has gravitational energy and momentum. According to the prevailing view in general relativity the components of the metric tensor are the relativistic generalization of the gravitational potential. The nonlinear terms in Einstein's equations are the squares of their partial derivatives, so the energy density of the gravitational field turns out to be quadratic in the gravitational field strength just like the energy density of the electromagnetic field is quadratic in the electric and the magnetic fields.

Identifying the gravitational field with the components of the metric tensor seems justified only in the limiting case when general relativity is compared with the Newtonian gravitational theory in order to determine what in general relativity (in that limiting case) corresponds to the gravitational potential and force. However, such an identification in general relativity itself is more than problematic. There is no tensorial measure of the gravitational field in general relativity since it can be always transformed away in the local inertial frame [3, p. 221]. This is problematic because if the gravitational field existed, then as something real it should be represented by a proper *tensorial* expression. For this reason not all relativists are happy with the identification of the components of the metric tensor with the gravitational field. Synge's view on this is well known – he insisted that the gravitational field should be modelled by "the Riemann tensor, for it *is* the gravitational field – if it vanishes, and only then, there is no field" [2, p. viii]. When gravitational phenomena are properly modelled by the spacetime curvature, which as something real is represented by the Riemann curvature tensor, it follows that gravitation (the spacetime curvature) makes no contribution to its own source – Einstein's equations are linear in the Ricci curvature tensor (the contraction of the Riemann curvature tensor) and the scalar spacetime curvature (the contraction of the Ricci curvature tensor). So, when gravitational phenomena are adequately modelled by the spacetime curvature it is evident that the gravitational field is not something physically real, that is, it is not a physical entity. It is a *geometric* field at best and as such does not possess any energy and momentum.

According to the second view there is indirect astrophysical evidence for the existence of gravitational energy. That evidence is believed to come from the interpretation of the decrease of the orbital period of a binary pulsar system, notably the system PSR 1913+16 discovered by Hulse and Taylor in 1974 [10]. According to that interpretation the decrease of the orbital period of such binary systems is caused by the loss of energy due to gravitational waves emitted by the systems. Almost without being challenged (with only few exceptions [11, 12, 13]) this view holds that the radiation of gravitational energy from the binary systems, which is carried away by gravitational waves, has been indirectly experimentally confirmed to such an extent that even the quadrupole nature of gravitational radiation has been also indirectly confirmed.

It may sound heretical, but the assumption that the orbital motion of the neutron stars in the PSR 1913+16 system loses energy by emission of gravitational waves *contradicts general relativity*, particularly the geodesic hypothesis and the experimental evidence which confirmed it. The reason is that by the geodesic hypothesis the neutron stars, whose worldlines are *geodesics*[5], *move by inertia without losing energy* since the very essence of inertial motion is motion without any loss of energy. Therefore no energy is carried away by the gravitational waves emitted by the binary pulsar system. So the experimental fact of the decay of the orbital motion of PSR 1913+16 (the shrinking of the stars' orbits) does not constitute evidence for the existence of gravitational energy. That fact may most probably be explained in terms of tidal friction as suggested in 1976 [15] as an alternative to the explanation given by Hulse and Taylor.

Despite that there is no room for gravitational energy in general relativity, it is an experimental fact that energy participates in gravitational phenomena, but that energy is well accommodated in the theory. Take for example the energy of oceanic tides which is transformed into electrical energy in tidal power stations. The tidal energy is part of gravitational phenomena, but is not gravitational energy. It seems most appropriate to call it *inertial energy* because it originates from the work done by inertial forces acting on the blades of the tidal turbines – the blades further deviate the volumes of water from following their geodesic (inertial) paths (the water volumes are already deviated since they are prevented from falling) and the water volumes *resist* the further change in their inertial motion; that is, the water volumes exert *inertial* forces on the blades. With respect to the resistance, this example is equivalent to the situation in hydroelectric power plants where water falls on the turbine blades from a height (the latter example is even clearer) – the blades prevent the water from falling (i.e. from moving by inertia) and it resists that change. It is that resistance force (i.e. inertial force) that moves the turbine, which converts the inertial energy of the falling water into electrical energy. According to the standard explanation it

---

[5]The neutron stars in the PSR 1913+16 system had been "modelled dynamically as a pair of orbiting point masses" [14], which means that (i) the tidal effects had been ignored and (ii) the worldlines of the neutron stars *as point masses* had been in fact regarded as exact geodesics.

is the kinetic energy of the falling water (originating from its potential energy) that is converted into electrical energy. However, it is evident that behind the kinetic energy of the moving water is its inertia (its resistance to its being prevented from falling) – it is the inertial force with which the water acts on the turbine blades when prevented from falling. And it can be immediately seen that the inertial energy of the falling water (the work done by the inertial force on the turbine blades) is equal to its kinetic energy (see *Appendix C: On inertial forces, inertial energy and the origin of inertia*).

Conclusion

The fact that for decades the efforts of so many brilliant physicists to create a quantum theory of gravity have not been successful seems to indicate that those efforts might not have been in the right direction. In such desperate situations in fundamental physics all options should be on the research table, including the option that quantum gravity as quantization of gravitational interaction is impossible because a rigorous treatment of gravity as a manifestation of the non-Euclidean geometry of spacetime demonstrates that there is no gravitational interaction and therefore there is nothing to quantize.

References

[1] J. Murugan, A. Weltman, G.F.R. Ellis (eds.), *Foundations of Space and Time: Reflections on Quantum Gravity* (Cambridge University Press, Cambridge 2011); B. Boo-Bavnbek, G. Esposito, M. Lesch (eds.), *New Paths Towards Quantum Gravity* (Springer, Berlin Heidelberg 2010); D. Oriti (ed.), *Approaches to Quantum Gravity: Toward a New Understanding of Space, Time and Matter* (Cambridge University Press, Cambridge 2009)

[2] J. L. Synge, *Relativity: the general theory.* (Nord-Holand, Amsterdam 1960)

[3] W. Rindler, *Relativity: Special, General, and Cosmological* (Oxford University Press, Oxford 2001)

[4] W. de Sitter, Over de relativiteit der traagheid: Beschouingen naar aanleiding van Einsteins hypothese, *Koninklijke Akademie van Wetenschappen te Amsterdam* **25** (1917) pp 1268–1276

[5] J.B. Hartle, *Gravity: An Introduction to Einstein's General Relativity* (Addison Wesley, San Francisco 2003) p 169

[6] H. Minkowski, Space and Time. New translation in: H. Minkowski, *Space and Time: Minkowski's Papers on Relativity* (Minkowski Institute Press, Montreal 2012) pp. 111–125 (http://minkowskiinstitute.org/mip/); included in this book as Appendix A

[7] V. Petkov, *Relativity and the Nature of Spacetime*, 2nd ed. (Springer, Heidelberg 2009)

[8] V. Petkov, On the Reality of Minkowski Space, *Foundations of Physics* **37** (2007) 1499–1502

[9] A. Pais, *Subtle Is the Lord: The Science and the Life of Albert Einstein* (Oxford University Press, Oxford 2005) p 179

[10] R.A. Hulse, J.H. Taylor, Discovery of a pulsar in a binary system, *Astrophys. J.* **195** (1975) L51–L53

[11] N. Rosen, Does Gravitational Radiation Exist? *General Relativity and Gravitation* **10** (1979) pp 351–364

[12] F.I. Cooperstock, Does a Dynamical System Lose Energy by Emitting Gravitational Waves? *Mod. Phys. Lett.* **A14** (1999) 1531

[13] F.I. Cooperstock, The Role of Energy and a New Approach to Gravitational Waves in General Relativity, *Annals of Physics* **282** (2000) 115–137

[14] J.H. Taylor, J.M. Weisberg, Further experimental tests of relativistic gravity using the binary pulsar PSR 1913+16, *Astrophysical Journal* **345** (1989) 434–450

[15] S.A. Balbus and K. Brecher, Tidal friction in the binary pulsar system PSR 1913+16, *Astrophysical Journal* **203** (1976) pt. 1, 202–205

Index

About the author

Vesselin Petkov received a graduate degree in physics from Sofia University, a doctorate in philosophy from the Institute for Philosophical Research of the Bulgarian Academy of Sciences, and a doctorate in physics from Concordia University in Montreal. He taught at Sofia University and Concordia University.

He is one of the founding members of the *Institute for Foundational Studies "Hermann Minkowski"* (minkowskiinstitute.org) whose most distinct feature is the employment of a research strategy based on the successful methods behind the greatest discoveries in physics. In this sense the *Minkowski Institute* is without a counterpart in the world.

This book, and especially the paper "Can Gravity be Quantized" (included as Appendix D) is intended to provide an idea of how such a strategy can identify and examine rigorously even (at first sight) heretical research directions.

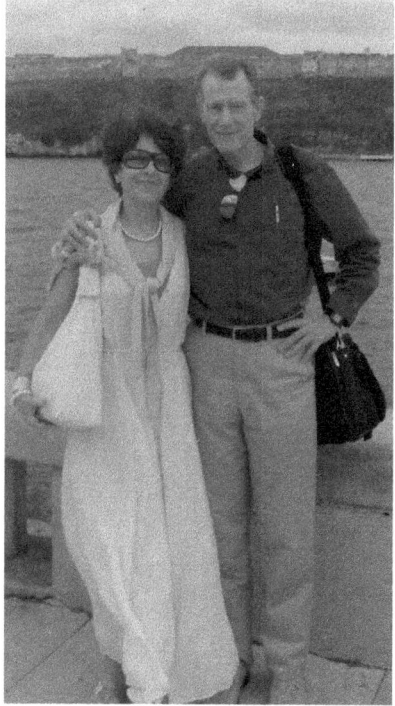

Vesselin with his wife Svetla in Havana, Cuba in January 2011

www.ingramcontent.com/pod-product-compliance
Lightning Source LLC
Chambersburg PA
CBHW071904200326
41519CB00016B/4508